INTRODUCTION TO HYDROGEOLOGY

UNESCO-IHE LECTURE NOTE SERIES

Introduction to Hydrogeology

JOHANNES C. NONNER

Second edition

CRC Press
Taylor & Francis Group
Boca Raton   London   New York   Leiden

CRC Press is an imprint of the
Taylor & Francis Group, an **informa** business

A BALKEMA BOOK

Library of Congress Cataloging-in-Publication Data

Nonner, Johannes C., 1948–
    Introduction to hydrogeology / Johannes C. Nonner.—2nd ed.
      p. cm.
    Includes bibliographical references.
    ISBN 978-0-415-87555-4 (pbk. : alk. paper) 1. Hydrogeology.
      I. Title.

GB1003.2.N65 2010
551.49—dc22
                                                     2009036990

Published by: Taylor & Francis/Balkema
P.O. Box 447, 2300 AK Leiden, The Netherlands
e-mail: Pub.NL@tandf.co.uk
www.balkema.nl, www.tandf.co.uk, www.crcpress.com

ISBN 978-0-415-87555-4 (paperback edition)

Contents

Foreword

The Hydrology and Water Resources training programme offered at
the International Institute for Infrastructural and Environmental Engineering (IHE) in The Netherlands also specialises in groundwater. As
a contributor to this programme I had the task to collect instruction
material on groundwater which could be used for the training of participants. Although excellent textbooks on groundwater are available, I
could not swiftly identify an introductory textbook on this subject and
therefore decided to prepare 'Lecture Notes'. These notes had been
used in groundwater classes until the opportunity arose, in 1998, to
convert the lecture material into a textbook. The present 'Introduction
to Hydrogeology' is the result of this conversion. This opportunity
and my wish to assist in passing on current knowledge on groundwater have motivated me to make the time available to compile this
textbook.

In this textbook a systems approach is followed. This means that the
groundwater system has been considered as the framework for the compilation of this Introduction. In Chapter 1, following a historical introduction, the groundwater system is placed within the earth hydrological
cycle. Brief introductions are given to the other hydrological systems
including the atmospheric system, the surface water system and the
unsaturated zone. Chapter 2 describes basic properties of 'water and
rock' in the groundwater system and introduces 'hydrogeological classifications'. The chapter concludes with an outline on the occurrence
of groundwater in various rock types and field investigation methods.
Chapter 3 is devoted to groundwater flow. In this chapter the emphasis
is on the evaluation of regional flow in groundwater systems, rather
than on local groundwater in the vicinity of infrastructural works.

In Chapter 4, groundwater balances are discussed in general terms,
but also in relation to the various rock types and climatological conditions on earth. Chapter 5 focuses on groundwater chemistry. The main
processes are discussed and the chemistry of groundwater in selected
rock types is evaluated. The final Chapter 6 has the title 'Development of
Groundwater'. Practical topics are dealt with in this text and the chapter
includes an extensive outline on the selection of field investigations and

guidelines for the determination of groundwater availability using a groundwater balance approach. For personal training in various aspects of the 'science of hydrogeology', a set of groundwater problems has been added.

I would like to thank all persons who have, in some way or another, contributed to this textbook. In the first place I owe a lot to my parents for giving me the opportunity to undertake my studies in Utrecht and Amsterdam, and to my wife and children for their moral support. I would also like to thank Prof. Dr. Ir. J.C. van Dam, Prof. Dr. J.J de Vries, Dr. C.A.J. Appelo and Dr. P.J.M. de Laat for their critical reading of the draft versions. Their comments and suggestions have been very helpful. Finally, I would like to express my gratitude to Hans Emeis for his assistance in the preparation of the drawings, and to staff and participants of IHE who have also made a contribution to this textbook.

In the first revision of the book valuable suggestions by external reviewers were taken into account. This meant that more attention was paid to the processes in the unsaturated zone, including groundwater recharge. The investigation methods were outlined in the chapters where the relevant theory is discussed and not in the chapter on groundwater development. The 'References and Bibliography' section was also extended, some figures were improved and the unevitable 'typing errors' were corrected as well.

In the second edition comments ventilated in late reviews of the book have been incorporated. Notably, the chapter on groundwater chemistry has been expanded. In addition, some text and figures have been improved.

Johannes C. Nonner

CHAPTER 1

Introduction

1.1 GENERAL BACKGROUND

1.1.1 *The scope of the science*

Definition of hydrogeology

Hydrogeology can be considered as one of the earth-sciences that has gained in popularity in recent years. Occasionally there has been confusion with respect to the precise definition of hydrogeology. Although the term hydrogeology was already used in the 19th century, it was only at the start of the 20th century that the scientist Mead (1919) gave the term a wider meaning. He defined hydrogeology as 'the study of the occurrence and movement of subterranean waters'. Meinzer (1923) used the term geohydrology to describe in principle the same physical science.

The basic definition of hydrogeology as formulated by Mead is a very useful definition. In this textbook, however, a slightly different formulation will be used. Hydrogeology is 'the study of the occurrence, movement and chemistry of groundwater in its geological environment'. This definition emphasises the role that the science of geology plays in the study of groundwater. A sound knowledge of the geology of an area should form the basis for all hydrogeological assessments.

Relation with other sciences

Hydrogeology is related to other sciences, as shown in Figure 1.1. There are first relations with basic sciences like mathematics, physics, and chemistry. Knowledge of these subjects is indispensable for a full understanding of hydrogeology. Other earth-sciences that need to be mastered by the hydrogeologist are geology, surface water hydrology and meteorology. It should be said, however, that no scientist can master all these subjects in detail and therefore much of the research done in the field of hydrogeology has been accomplished by joint efforts between hydrogeologists, geologists, hydrologists, and meteorologists.

Worth mentioning is also the relation between hydrogeology and engineering. Hydrogeology plays an important role in the design work that engineers undertake for the benefit and welfare of mankind. The hydrogeologist can supply information on groundwater tables,

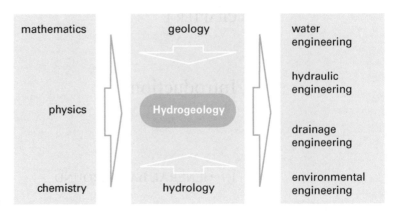

Figure 1.1. Hydrogeology and its relation with other sciences.

available groundwater resources and water quality aspects that the engineer wants to know for the design of, for example, public and domestic water supplies, irrigation schemes, flood protection works, and even office and industrial buildings.

1.1.2 *The application of hydrogeology*

Engineering applications   The science of hydrogeology is applied in various fields of engineering. The most important fields of civil and agricultural engineering using hydrogeological assessments can be summarised as follows:
– Water engineering; i.e. the field of engineering that deals with the supply of water for domestic and industrial use, and irrigation.
– Hydraulic engineering; i.e. the engineering that occupies itself with the construction of water works.
– Drainage engineering; i.e. the engineering branch that is involved in the drainage of lowland areas and the dewatering of mines.
– Environmental engineering; i.e. the engineering that deals with the conservation of the natural environment and the creation of new nature.

Water engineering   Water engineering works include water wells and well fields whereby the assessment of well locations, well depths and water quality is important. Wells supply groundwater to user areas to satisfy water demand. In recent decennia, the use and demand have increased considerably. The trend is visible in many countries in the developing as well as in the developed world. For example, in the United States groundwater consumption was growing at an even faster rate than the population. This is illustrated in Figure 1.2 showing that the groundwater supplies in urban areas, and for irrigation were largely responsible for the fast growth of groundwater consumption in recent times.

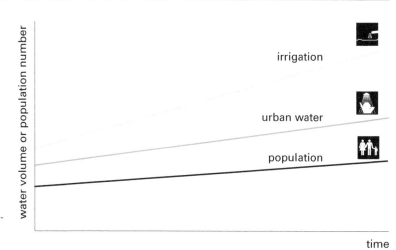

Figure 1.2. Increase in ground-water consumption versus the increase in population.

It is not surprising that in the developed world groundwater supplies in urban areas and for irrigation have increased so rapidly. The trend that people move away from rural areas and settle in densely populated urban areas partly explains the increase in water use as delivered by groundwater supply schemes. Also, the increased use of water for personal hygiene (showers, washing machines) and the use of water for the sprinkling of lawns and gardens has put a strain on the water supply schemes in urban areas. In the agricultural sector farmers are increasingly developing areas where crops heavily depend on irrigation from groundwater.

In many of the developing countries one can observe similar trends in groundwater consumption. Also in these countries there is a growing urbanisation and awareness of personal hygiene that has led to additional claims on available groundwater resources. Since it is well known that there is a rapid increase in groundwater irrigation schemes in large parts of the developing world, similar claims on groundwater can be expected from the irrigation sector. This phenomenon is not only visible in arid countries, but also in the humid tropics where irrigation schemes are implemented to secure the harvest of two or three crops per year.

Hydraulic engineering   Water works or water-associated works that are designed by hydraulic and civil engineers include diversion weirs in rivers, dams at reservoirs, dikes and embankments along rivers, road supports and foundations for buildings. Taking into account the increasing complexity of these structures and, on the other hand, the tendency to minimise the risk of failure one should be aware that hydrogeology plays an increasingly important role in this field of engineering. For example, the spatial distribution and fluctuations in groundwater tables (see section 1.2.4) in an area are determining factors for the stability computations carried out for the design of water works.

Drainage engineering

Dewatering and drainage infrastructures are developed by engineering geologists, mining experts, and agricultural engineers. For engineering geologists and mining experts hydrogeological assessments are valuable when they compute pump capacity to be installed in building pits, and open cast or underground mines. Positioning of these pumps at the correct locations in the pits or mines is essential. Knowledge of hydrogeology is also crucial to the agricultural engineer who wishes to install horizontal drainage systems or vertical drainage schemes consisting of wells. For example, in The Netherlands the horizontal drainage system in the low-lying part of the country ('the polders') has been designed on the basis of a thorough hydrogeological study. The system that has been installed is capable of draining away the excess shallow flow of groundwater, while largely preventing the upconing of brackish water.

Environmental engineering

Environmental engineering concerns the setting up of measures and the organisation of operations to prevent and restore adverse effects on groundwater. In particular in the past decennium, there has been full agreement amongst scientists that groundwater is a precious resource which has to be carefully assessed and managed. Not only has this been the case in the minds of scientists, but also in the minds of a much wider public. There are two principal causes for environmental concern related to groundwater resources. First, there is the concern for the decline in groundwater tables and depletion of resources as a result of water engineering activities including the installation of wells and well fields. Secondly, there is the concern for a deterioration in groundwater quality caused by human negligence at waste disposal sites, the use of toxic chemicals in agriculture, salt water intrusion or, simply, by a lack of sanitary protection.

Increasing role of hydrogeology

One conclusion that can be drawn from the above is that the science of hydrogeology is playing a central role in many engineering disciplines. However, it will be clear that not only hydrogeologists and engineers focus on hydrogeology. Increasingly so, other professionals including 'politicians and other decision makers', environmentalists, biologists and even economists have to pay attention to groundwater issues. Proper training in hydrogeological assessments is also an essential part of the game. Figure 1.3 presents a photo showing a group of students to be trained in taking observations at a rock outcrop and conducting geological measurements. Groundwater can not be taken for granted anymore. People will have to pay a larger price for the proper

Figure 1.3. Hydrogeological assessments often start with rock observations. Here, students are inspecting slates and quartzites, dissected by joints and faults.

management and rehabilitation of groundwater resources in the foreseeable faults.

1.1.3 *Historical background of the science*

Development of the science

In the old Greek and Roman civilisations philosophers and scientists have speculated about 'underground waters'. They were puzzled by springs, and the discharge, of water in rivers in dry periods, long after precipitation had stopped. The Greek Plato (427-347 B.C.) and other philosophers offered a solution. They thought that groundwater originated from caverns which were connected to the ocean. By the action of waves, water of the sea was transported from the ocean upward into caverns and from there to springs and rivers. They assumed that underground purification filtered away the salts in seawater, in order to explain that spring water and river water are fresh.

The true explanation for the occurrence of springs and discharges in rivers during dry periods was put forward by the French scientists Perrault (1608-1680) and Mariotte (1620-1684). Perrault suggested that the amount of precipitation in the Seine catchment in France could easily account for all the discharge in the river Seine. Mariotte carried out infiltration experiments in the catchment and discharge measurements in the Seine. He found that precipitation could infiltrate into the ground and underlying rocks in appreciable quantities. Mariotte concluded that these rocks, acting as a storage medium, could sustain springs and a year around discharge in the Seine.

In the 19th century another development drew wide attention from other scientists in the field of hydrogeology. The French water engineer Darcy (1803-1858) made a start with groundwater hydraulics (Darcy, 1856). He did experiments with sand columns and formulated his well-known formula to compute the 'flow rate' of groundwater in porous rocks.

Flow formulae and modelling

Since the 19th century many scientists including the mentioned Mead and Meinzer (see section 1.1.1) have made respectful contributions to the science of hydrogeology. Modern trends in hydrogeology concern the development of flow formulae and modelling, the introduction of the 'flow systems' concept (see below and section 3.3), and the focus on hydrochemistry and groundwater contamination.

Flow formulae based on 'Darcy's Law' have been developed. Scientists including Thiem (1906), Theis (1935) and Jacob (1950) developed formulae for radial flow to wells. The formulae concern the flow of fresh groundwater with constant density. The range of formulae includes expressions describing the flow to wells in various natural environments. Flow formulae for open- and closed 'groundwater systems' (see section 1.2.5), for 'soft and hard' rocks and for steady and non-steady conditions were developed.

Other types of formulae were outlined at a later stage, for example, those dealing with the flow of groundwater in an environment where water density varies from place to place. In this respect one may think of a phenomenon that can be observed in many coastal areas: the flow of groundwater in fresh water lenses which overlay saline groundwater. Scientists including Van Dam (1983) developed equations for the flow in these fresh water lenses and the position of the fresh-saline interface taking into account various stresses acting on the groundwater system.

The scientific efforts to develop appropriate formulae for the flow to wells and for the flow in areas with a varying water density were initially based on an analytical approach; i.e. on the application of analytical methods to solve the general groundwater flow equations. From about 1960 less attention has been paid to the development of analytical methods. Models based on numerical methods became rapidly popular and they form the basis for most of the flow computations presently carried out. Compared with the flow equations based on analytical methods, models are much better suited to describe ground-water flow in areas where the groundwater system is of a complex nature. Also through the introduction of personal computers, the use of models increased quickly. The ability to process the computations fast is a most essential asset of the modern brand of computers.

Figure 1.4, for example, presents an application of a modern flow model. The diagrams show the model-calculated paths of groundwater flow between galleries for 'artificial recharge' (see section 1.2.5 and 6.3) and a well field in the complex sandy groundwater system in the eastern part of The Netherlands. The model was built to assess the scope for the restoration of groundwater tables in a sensitive area using artificial recharge systems.

Flow systems analysis   A challenging recent development in the science of hydrogeology has been the introduction of the flow systems concept. Traditionally hydrogeo-logical classifications were mainly based on the rock types in a particular area. In the flow systems concept as formulated by Toth (1962), classifications were suggested which are based, not only on rock types,

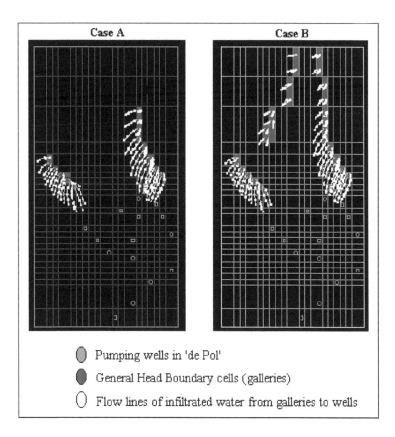

Figure 1.4. Computer printout showing flowlines from artificial recharge galleries to pumping wells in the complex sandy groundwater system in the de Pol area, Doetinchem, The Netherlands. Case A: two galleries operational, Case B: five galleries operational (Abushaar, 1997).

but also take into account groundwater flow distributions. For example, this concept has been applied in The Netherlands whereby an atlas has been prepared showing the flow systems in various parts of the country. The flow systems concept provides a flexible approach towards groundwater analyses, in particular for areas where human interferences in the groundwater system have to be studied.

Groundwater chemistry and groundwater contamination

Other recent developments in the science of hydrogeology include the work done on groundwater chemistry and groundwater contamination. In the field of groundwater chemistry many scientists including Appelo and Postma (2009) have paid attention to the inter-relationship between the chemical constituents dissolved in groundwater and the chemical composition of the associated rock types. Work leading to a classification of groundwater in terms of 'water types' (see section 5.1) was also initiated and developed by Stuyfzand (1999). Research has also increased in the field of groundwater tracing in fractured and karstic rock areas. In this field, the behaviour of chemical components is studied with the objective to assess the flowpaths of groundwater through rocks such as basalts, rhyolites, chalks, limestones and dolomites.

As a result of the growing concern of groundwater contamination much work has been done in analysing solute transport in groundwater. Bear and Verruijt (1987), and Kinzelbach (1986) are amongst the scientists who gave impetus to these developments. The phenomenon of dispersion takes a central place in their research, while physicochemical processes such as adsorption and the decay of chemical components have also been considered. Perhaps the most recent topic is the coupling of physical and physico-chemical processes in solute transport modelling.

1.1.4 *History of groundwater use*

The early history

Since the early days of humanity people have used groundwater for domestic consumption and for irrigation. Springs were utilised for these purposes, but also wells were constructed to tap groundwater resources. Centuries ago, well construction practices were first carried out during the old civilisations in China, in the Middle East and in Egypt. Due to the scarcity of water in these areas a lot of effort was applied in constructing these wells which in some cases reached amazing depths. Large diameter wells were often dug to depths over 100 m using primitive equipment. Drilled wells that were installed with augering equipment could reach depths even in excess of 500 m.

An amazing achievement in groundwater development concerned the construction of so-called 'khanats' that were first installed in Iran more than 2500 years ago. Later on, the art of constructing khanats was also practised in Afghanistan and in Egypt. Figure 1.5 shows a

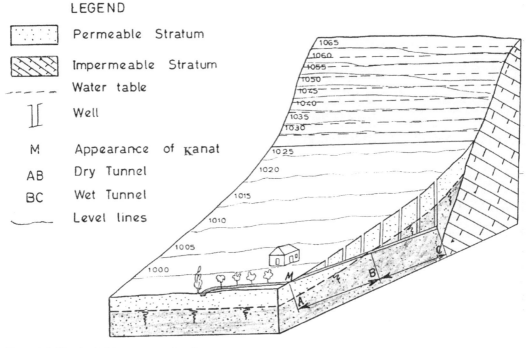

LEGEND

▭	Permeable Stratum
▨	Impermeable Stratum
- - - -	Water table
⫪	Well
M	Appearance of ĸanat
AB	Dry Tunnel
BC	Wet Tunnel
‿	Level lines

Figure 1.5. Sketch of a khanat system from Iran.

rough sketch of a khanat system. Khanat systems are composed of series of large diameter wells that were dug into sedimentary rocks and alluvial fans. The wells were connected by horizontal galleries (tunnels) reinforced by natural rock. Khanat systems tap the groundwater table and the intercepted water was, and still is, used for domestic supplies and small irrigation schemes.

Recent developments   Since the 12th century, the construction of wells to provide groundwater became a popular activity in Europe. In particular the successful installation of a number of free flowing 'artesian' wells in Belgium, England and Italy stimulated the use of drilling techniques to construct wells to guarantee domestic and irrigation water for the local population. Traditional auger drilling and percussion drilling techniques were upgraded. At the end of the 19th century an important break-through was the development of the rotary drilling technique (see also section 2.3.2). In rotary drilling which was found to be much faster than conventional drilling, drillers for the first time used heavy bentonite mud to guarantee hole stability during drilling. The installation of temporary casing was not longer necessary. In the 20th century well drilling techniques have been further developed.

Reverse rotary drilling methods, air-rotary percussion methods, and geophysical well logging techniques were introduced.

1.2 HYDROLOGICAL SYSTEMS

1.2.1 *Earth and water*

Oceans, atmosphere and continents

Hydrogeology is defined as the study of the 'occurrence, movement and chemistry of groundwater in its geological environment' (see section 1.1). When one studies hydrogeology one has to pay attention to the processes that play a role in the oceans, in the atmosphere and at the continents of the world. There is a rather indirect, but nevertheless significant, relationship with the oceans and the atmosphere. Obviously the science of hydrogeology is directly related to the continents. The occurrence and movement of groundwater very much depends on the structure of the rocks contained in the continents and the morphological and climatic conditions occurring at or near the surface of the continents.

One can take a brief look at the properties of the oceans, atmosphere and continents respectively. Table 1.1 shows that the oceans and seas have a surface area of $361*10^6$ km^2, making up about 75% of the global surface. The other part is occupied by the landmass of the five continents. The largest part of the oceans is concentrated south of the equator. The depths of the oceans varies from less than 10 m in peripheral seas to over 11000 m in longitudinal troughs which are mainly located in the Pacific Ocean. The total volume of water in the oceans and seas amounts to about $1.4*10^9$ km^3 which is by far the largest water volume on the earth.

The atmosphere can be subdivided into various atmospheric layers. Most of the atmospheric water is contained in the bottom layer: the troposphere. The temperature in the troposphere, which has a thickness of about 12 km, decreases with an increase in height. The layer above the troposphere is called the stratosphere. Surprisingly, in this

Table 1.1. Main world water quantities (Lvovich, 1979).

Item	Area (10^6 km^2)	Volume (10^6 km^3)	Percentage (%)
Oceans and seas	361	1370	94.2
Icecaps and glaciers	16.2	24	1.65
Atmospheric water	504	0.01	< 0.01
Lakes and reservoirs	1.55	0.13	< 0.01
River channels	< 0.1	< 0.01	< 0.01
Groundwater	130	60	4.13

atmospheric layer the temperature tends to increase with height. Nearly all the atmospheric water is present in the form of gaseous water vapour. Table 1.1 gives an idea of the amount of water vapour in the atmosphere. The computed equivalent amount of water amounts to only $0.01*10^6$ km^3. It can be concluded from the table that the amount of water in the atmosphere is very small as compared with the amounts of water 'stored' in the oceans and seas, and on the continents.

At continents features of interest can be observed at land surface. These features include lakes, reservoirs, streams and rivers. Their presence and their size depend on the climate that prevails in the area in which one is interested. For example, the rivers with the highest discharge can be expected in tropical areas where precipitation is highest. No rivers at all or rivers that only flow intermittently are typical for arid areas where precipitation is not abundant. The amount of water stored in lakes, reservoirs, streams and rivers is comparatively small and amounts to a mere $0.13*10^6$ km^3.

Features below land surface include 'hard and soft' rocks, molten magma, and other liquid material. For a general orientation one may first have a look at Figure 1.6, presenting the various interior components of the earth. The centre space of the earth is occupied by the so-called 'earth core' which consists of iron and nickel compounds. The 'mantle' with a thickness of about 3000 km surrounds the core of the earth and is made up of ultra-basic rocks with a density in the order of 3000 to 4000 kg/m^3. The earth 'crust' is the outermost shell of the earth. With its thickness varying between 12 km below the oceans to 35 km below the continents the rock density in the crust varies between 2500 and 3000 kg/m^3.

The upper part of the crust at the continents contains zones with water that is generally referred to as groundwater. The types of rocks contained in the continents give an indication of the thickness of

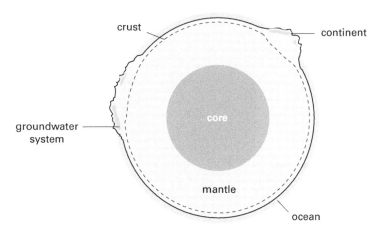

Figure 1.6. Cross section through the earth showing groundwater systems at the continents.

these groundwater systems (see section 1.2.5). On the average ground-water systems have a thickness between 10 m and 500 m. However, in some areas including coastal plains these systems may be more than 1000 m thick. Table 1.1 shows that on earth the total volume of water stored in groundwater systems is in the order of $60*10^6$ km^3. The table indicates that, disregarding the icecaps and glaciers, most of the fresh water reserves of the earth are present in the form of groundwater.

The hydrological cycle

The volumes of water contained in the various environments of the earth, oceans, atmosphere and continents, have been described above. As the French hydrogeologists Perrault and Mariotte already found out, this water can be transported from one environment to the other. The general process is presented in Figure 1.7. Water evaporates at the oceans and is transported as water vapour in the atmosphere to the continents. Above the continents the water vapour is transformed into precipitation in the form of rain, hail or snow which fall on the surface of the earth. Rain, molten hail and snow may be directed towards lakes, streams or rivers or alternatively infiltrate into the ground, where they recharge groundwater systems. From lakes, streams and rivers and from the groundwater systems the water flows back into the oceans and seas. In the oceans and seas water evaporates again, is transported to the continents, etc. The process described is the transport of water in a cyclic process; it is also referred to as 'the hydrological cycle'.

The process outlined above is a very generalised one and in reality it is far more complex. For example, not all the evaporated water from the oceans will be transported to the continents as water vapour, but part of it will return directly to the ocean in the form of precipitation. Also, not all the rain, hail and snow that falls on the earth surface will infiltrate into the ground or be transported to lakes, streams and rivers.

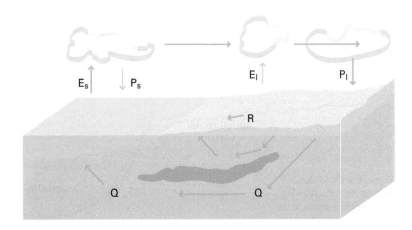

Figure 1.7. Schematic representation of the hydrological cycle. (See equations (1.1) and (1.2) for explanation of symbols)

A substantial portion of this water will evaporate from land surface or the soil. In addition, not all the water that recharges the groundwater system will directly flow back to the seas and oceans. The scientists Perrault and Mariotte already concluded that groundwater may also flow from the groundwater system into a river.

Table 1.2 shows the inflows and outflows of water or water vapour associated with the oceans and seas, and with the continents. It can be deduced from Table 1.2 that an amount of water in the order of $0.43*10^6$ km^3/year evaporates from the oceans and seas. Less than 10% of this amount, in the order of $0.036*10^6$ km^3/year is transported as water vapour to the continents. On the continents themselves this water vapour, and the water evaporated from land surface comes down as precipitation to the amount of about $0.11*10^6$ km^3/year. The amount of water which flows back from lakes, streams and rivers, and groundwater systems to the oceans and seas is equivalent to the amount of water vapour transported to the continents: $0.036*10^6$ km^3/year.

Figure 1.8 shows one of the major rivers on earth transporting water back to the ocean. The river Ganges portrayed in the picture collects water in the Himalayas and takes the water to the Indian Ocean. Nowadays, far lesser amounts of water are flowing through the Ganges to the ocean, as a result of the diversion of river water for domestic use and irrigation schemes.

The inflows and outflows as outlined in Table 1.2 can also be seen as the terms of so-called 'water balances'. One can have a closer look at the concept of water balances. Water balances can be set up for predefined environments such as the oceans and seas, the continents and even lake, stream and river systems or groundwater systems. A water balance essentially relates the flows of water into an environment to the flows of water leaving this environment. In a state of equilibrium the inflow of water is equal to the outflow of water. In a state of non-equilibrium the difference between inflow and outflow equals the change in water storage. For equilibrium conditions one can formulate

Table 1.2. Flow volumes of water at the globe (Shiklomanov, 1997).

Item	Flow (10^6 km^3/year)	Flow (mm/year)
Oceans and seas		
Evaporation	0.434 (out)	1202 (out)
Precipitation	0.398 (in)	1103 (in)
Inflow from groundwater and rivers	0.036 (in)	
Continents		
Evaporation	0.071 (out)	476 (out)
Precipitation	0.107 (in)	718 (in)
Outflow via groundwater and rivers	0.036 (out)	(out)

Figure 1.8. View of the majestic Ganges River near Rishikesh, India.

the general water balance for the oceans and seas. Considering the water balance terms listed in Table 1.2 one can set up the following expression for the oceans and seas:

$$P_s + (R + Q) = E_s \tag{1.1}$$

For the continents one could formulate:

$$P_1 = (R + Q) + E_1 \tag{1.2}$$

where:

P_s = precipitation at seas and oceans (km^3/year)
E_s = evaporation at seas and oceans (km^3/year)

P_1 = precipitation at continents (km³/year)
E_1 = evaporation at continents (km³/year)
$(R + Q)$ = groundwater and river outflow (discharge) to oceans and seas (km³/year)

Is the relationship above correct? One could think of other inflows and outflows of water related to oceans and seas, or to continents. For example, one could mention an inflow of water associated with magma originating from the earth mantle. However, it appears that these inflows are small. Alternatively one could imagine an outflow of water stored in oceans and seas, due to the entrapment of water in the sediments on the ocean floor. However, also these quantities are comparatively small or counter-balanced by other effects.

1.2.2 *Water at land surface*

Precipitation processes

Precipitation is an essential parameter to be considered for quite a number of hydrogeological assessments. It is therefore useful to get acquainted with the processes that cause precipitation. Precipitation including rainfall, hail or snow is formed around small nuclei in those parts of the atmosphere where water vapour reaches its saturation point. Three atmospheric processes including lateral moisture transport, convective moisture transport, and transport under the influence of mountain effects, further stimulate the formation of precipitation. The formation of precipitation caused by these processes can be described as follows:
i) In particular in the tropics, precipitation can be enormous when large masses of warm moist air move laterally into regions of colder air.
ii) Through convection warm moist air can move upward into higher situated layers of cold air and precipitation will follow.
iii) Due to the presence of mountain ranges air rich in water vapour will rise upward and precipitation will follow when the air reaches a colder environment.

Gathering precipitation data

How does one determine precipitation rates that may vary considerably from place to place and from time to time? For the proper collection of precipitation data, which are required to assess precipitation rates, one needs a relatively dense network of observation points. At these observation points cumulative or continuous measurement of precipitation can be carried out. For the daily or monthly cumulative recording of precipitation one can use standard rain gauges or pluviometers, and for continuous measurement of precipitation automatic recorders can be installed.

Figure 1.9 shows recently developed instrumentation for the measurement of precipitation. A computerised solid state recorder is attached to a rainfall collector. Using a pressure sensor, water levels

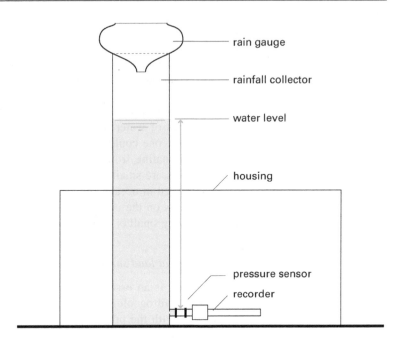

rain gauge

rainfall collector

water level

housing

pressure sensor

recorder

Figure 1.9. Rainfall collector
and recorder with pressure
sensor.

in the collector relating to the amounts of precipitation collected are
stored in a computerised memory chip placed in the recorder.

The processing of precipitation data can be carried out in various ways. For example, so-called 'isohyetal maps' can be prepared.
Isohyets which are contour lines connecting points of equal depth
of precipitation give a good insight in the aerial distribution of this
parameter. Figure 1.10 shows an isohyetal map for the catchment area
of the Fundres river in the Dolomites in Italy. The increase in daily
precipitation rate towards the higher areas is a result of the mountain
effects in the river catchment.

In addition, graphs of time series of precipitation rates are very useful. These graphs may present time series of average daily, monthly,
or yearly precipitation rates. However, graphs may also be prepared
showing precipitation frequency and precipitation intensities on an
hourly or even shorter time basis.

Table 1.3 has been prepared to obtain an impression of the precipitation rates for the continents. The data in this table are based on information supplied by Barry (1969) and Anon (1975). With the exception
of South America and Australia, it can be deduced from the table that
the average yearly precipitation on the continents is within close range
of each other; in the order of 600 to 700 mm/year.

*The concept of
evapotranspiration*

Evapotranspiration is another parameter that plays a role in hydrogeological assessments. The term evapotranspiration refers to two

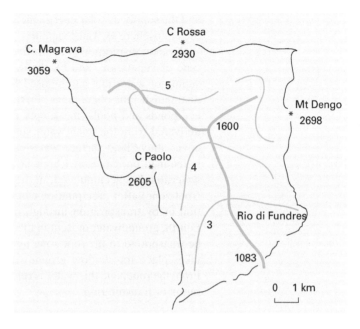

Figure 1.10. Map with isohyets for the mountainous Valle di Fundres area, Italy (Nonner, 1969).

3059   elevation in m above sea level

3   isohyet with precipitation in mm/day

Table 1.3. Precipitation, evaporation and runoff by continent.

Continent	Precipitation (mm/year)	Actual evapo-transpiration (mm/year)	Total runoff (mm/year)
Africa	670	510	160
Asia	610	390	220
Europe	600	360	240
N. America	670	400	270
S. America	1350	860	490
Australia/New Zealand	470	410	60

processes: it refers to evaporation and transpiration. Evaporation is the process whereby solar energy releases water molecules directly from free water surfaces. Water vapour is the result. The process occurs when the water molecules have absorbed sufficient energy to escape from the surface tension that holds them in the liquid or solid state. Transpiration is the process whereby the transformation of water into water vapour takes place through the vegetation at land surface. Water is taken up by the roots of plants and trees, and transported to their leaves. As a result of the physical properties of the vegetation itself and the supply of solar energy the water at the leaves can be freed into the atmosphere as water vapour.

Evapotranspiration is controlled by meteorological conditions, the type of vegetation, and by the supply of water. If one looks at the supply of water that can be used for evapotranspiration then one can distinguish four phenomena. First, at oceans, lakes, streams and rivers, ponds and pools, the supply of water for evaporation is directly delivered from the open water surface. Secondly, in the vegetational cover above land surface water can be supplied for evaporation that is retained on leaves and branches (intercepted precipitation). Third, water that has accumulated, in the upper part of the unsaturated zone (root zone) after precipitation can also be released by direct evaporation or by transpiration through plant roots (see also section 1.2.4). Fourth, groundwater present in the (saturated) groundwater system can be transported to the root zone by capillary flow. This transport only takes place for shallow groundwater conditions (see section 1.2.5). From the root zone this water is released into the atmosphere by evaporation or transpiration.

Figure 1.11. Picture of an area with high transpiration rates, also resulting from shallow saturated groundwater, Uden, The Netherlands.

Figure 1.11 shows an example of the influence of capillary flow and the resulting transpiration in the south of The Netherlands. The shallow groundwater is caused by the 'blockage' of the flow of groundwater against an impermeable fault. The groundwater transported to the root zones transpires in grassland and forest areas.

The supply of water that is available for evapotranspiration may vary from almost zero to flow rates in the order of several millimetres per day. For a given set of meteorological and land use (vegetation type) conditions, evapotranspiration rates reach maximum values if the supply of water is unlimited. This maximum rate is referred to as the 'potential evapotranspiration rate'. Potential rates will not always be reached if one considers evapotranspiration at land surface. Here, the amount of water available for evapotranspiration largely depends on the availability of water in the vegetational cover and the water content in the root zone. Water at land surface may be limited and this means that evapotranspiration rates do not reach the potential rates. One refers to these real rates as 'actual evapotranspiration rates'.

Evapotranspiration data collection

Can one determine evapotranspiration rates in an efficient way? Three of the best known methods to determine this parameter can be described as follows:

- *The pan evaporation method.* Here, the potential evaporation rate is directly measured as the loss of water from a large circular pan. For example, the United States Class A pan has a diameter of 1.21 m and a height of 0.25 m.
- *The lysimeter technique.* This technique involves the setting up of a cylinder, tank or basin filled with sediment having a typical vegetational cover. In these facilities which may have surface areas ranging from less than 1 m^2 to more than 1000 m^2, field measurement of water content, precipitation and discharge allow the compilation of water balances. From these water balances actual evapotranspiration rates can be computed.
- *The energy budget method.* This method is based on the measurement of meteorological parameters such as solar radiation, number of sunshine hours, humidity, windspeed etc. With formulae developed by Penman (1948) and Makkink and Van der Heemst (1965), potential evapotranspiration rates can be determined.

The processing of evapotranspiration data can be rather varied. The data may be presented on maps showing the areal distribution of potential or actual evapotranspiration rates, but even more popular are maps presenting the distribution of so-called 'precipitation surpluses'. These surpluses are the differences between precipitation and potential or actual evapotranspiration rates. Contour lines of equal precipitation surpluses may be drawn on the maps. Graphs showing time series of evapotranspiration rates or precipitation surpluses are also prepared to obtain an idea on the temporary fluctuations in these rates.

Table 1.3 also presents the actual evapotranspiration rates for the individual continents on earth. The table shows that these rates are in the order of 360 to 510 mm/year with the exception of South America. On this continent the high average precipitation of 1350 mm/year correlates with a high average evapotranspiration rate of 860 mm/year. Perhaps the most striking feature of the table is the fact that most of the precipitation on the continents is lost to evapotranspiration. Only a relatively small portion, in the order of 15% to 40%, flows towards rivers and lakes and recharges groundwater systems.

Infiltration and surface runoff

Infiltration and surface runoff are also important parameters for hydrogeological analyses. Infiltration can be defined as 'the portion of the precipitation that enters into the unsaturated zone' and surface runoff is 'that part of the precipitation that flows over land surface towards a surface water system' (see section 1.2.3). The first part of the precipitation that reaches land surface is retained in the vegetational cover and evaporates. If precipitation proceeds then water starts entering into the unsaturated zone and the amount of water stored in this zone increases. 'Recharge from precipitation' and groundwater flow may also be stimulated (see sections 1.2.4 and 1.2.5). However, if precipitation is continuing, then surplus water will start forming ponds and pools and eventually water flows over the land surface towards the surface water system.

The amount of surface runoff depends on the duration of precipitation, but also on precipitation intensity, the abundance and type of vegetation, the infiltration capacity of the unsaturated zone, and the slope

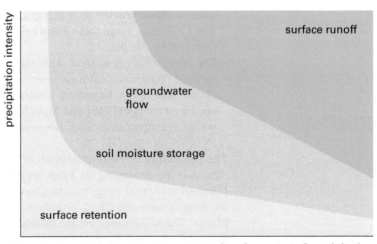

Figure 1.12. Distribution of precipitation (rainfall) after having reached land surface.

and roughness of the land surface. Figure 1.12 shows a diagram indicating that an increase in precipitation intensity decreases the portion of the precipitation that is retained in the vegetational cover (surface retention) and increases the part that is transformed into surface runoff. Sparse vegetation leads to small amounts of retained precipitation and also increases surface runoff. The infiltration capacity of the unsaturated zone has a profound effect on the amount of surface runoff (see section 1.2.3). A steep slope and limited roughness of the terrain decrease the 'resistance to flow' and may lead to a large surface runoff.

1.2.3 *Surface water systems*

Lakes, streams and rivers Surface runoff flows to a system of lakes, streams and rivers: the surface water system. The interaction between surface runoff, and the slope and the geology of an area determine the 'shape and pattern' of the surface water system. In particular the slope of the terrain determines the shape of streams and rivers. For example, in the upstream parts of a river catchment with moderate to steep slopes the shape of a river is relatively straight and the water is 'fast-flowing'. In the downstream parts where the land surface is usually flat, a winding or so-called 'meandering' river will be present and water will be moving at a much slower rate.

The prevailing geology in a catchment area largely determines the pattern of the surface water system. In a volcanic area, for example, scattered with volcanoes of more or less circular shape, the system will have a radial pattern. In areas with an alteration of 'soft' rocks such as shales and slates and 'hard' rocks like sandstones and quartzites the streams and rivers will mainly flow in the 'soft' rock and break only in places through the 'hard' rock. In such areas the surface water system usually has a rectangular appearance. In areas of a more or less homogeneous rock type the system has a so-called 'dendritic form'. The pattern resembles the branches and trunk of a tree.

Stream and river discharge Discharge can be defined as 'the volume of water passing through a cross section of a stream or river per unit of time'. Surface runoff taking place after persistent precipitation may contribute significantly to discharge. In a plot of discharge against time, the so-called 'hydrograph', the peaks in the graph represent this contribution. Another major contribution to stream or river discharge is the outflow of groundwater originating from recharge into the groundwater system (see section 1.2.5). This portion of the discharge is also called the 'baseflow' of a stream or river. The baseflow portion is represented by the bottom part of the hydrograph.

Figure 1.13 shows a small stream emerging from volcanic terrain in the Caribbean. The picture shows the baseflow of the stream originating

Figure 1.13. Officer, inspecting the base flow of a stream in densely vegetated terrain in the eastern part of Grenada.

from the volcanic groundwater system. In the rainy season the discharge of the stream may increase tremendously and cause flooding in the lower parts of the catchment.

The effect of precipitation on surface runoff and groundwater outflow as components of stream or river discharge can be further illustrated. For this purpose, one may consider the four diagrams presented in Figure 1.14, showing sketches of precipitation events and the shape of the corresponding hydrographs.

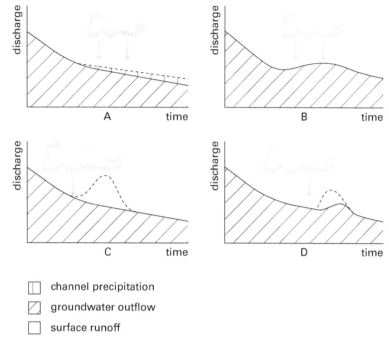

Figure 1.14. Hydrographs of river discharge in response to a precipitation (rain) event, indicated by the cloud and blue arrow.

☐ channel precipitation

▨ groundwater outflow

☐ surface runoff

A) The precipitation intensity is smaller than the infiltration capacity of the unsaturated zone and the volume of infiltrated water is smaller than the volume required to saturate this zone. This means that there is no surface runoff and no increased outflow from the groundwater system. In this case only the so-called 'channel precipitation' adds to the discharge in the stream or river.

B) The precipitation intensity is smaller than the infiltration capacity, but the volume of infiltrated water exceeds the water deficit in the unsaturated zone and recharge to the groundwater system takes place. This increases the groundwater flow in this system and will result in an increased contribution of groundwater outflow to the stream or river.

C) The precipitation intensity is larger than the infiltration capacity, and the volume of infiltrated water is smaller than the water deficit in the unsaturated zone. In this case mainly surface runoff adds to the discharge.

D) The precipitation intensity is larger than the infiltration capacity, and the volume of infiltrated water is larger than the water deficit in the unsaturated zone. Both surface runoff and groundwater outflow contribute to the increase in stream or river discharge.

Special cases are hydrographs of streams that do not continually show the groundwater outflow part. This usually implies that for part of the year there is no discharge at all. Only during short periods after heavy precipitation there are surface runoff events which are represented as isolated peaks in the hydrographs. These types of hydrographs may point to areas where the infiltration capacities are very low as a result of the impermeable nature of the rocks in the contributing areas. Alternatively, the hydrographs may have been compiled for catchments where the groundwater systems have no hydraulic contact with the surface water systems.

Field measurement of discharges

What can be said about the measurement of discharges in streams and rivers? Normally, discharges can be measured by stream gauging instruments equipped with propellers or lightweight cups. They measure the actual water velocity at various depths at verticals in a cross section. The equipment can be engaged by wading through the stream or river or it can be handled from a bridge or boat.

Related water levels in streams and rivers can be measured with staff gauges that are read by an observer at pre-set time intervals of say, an hour or a day. During extreme floods the levels can be estimated from flood marks in trees or at buildings along the stream or river. For a continuous registration of water levels, recorders may be used. These recorders may consist of mechanical devices working with clocks or they may contain computerised parts.

Figure 1.15 presents a front view of a clock-driven chart in a recorder connected to a float in a stilling well. The recorder was installed at the 'right' bank of the river Rhine, at the border between Germany and The Netherlands. Figure 1.16, on the other hand, shows the set up for a computerised solid state recorder installed in a pipe, driven in a river bottom. The chart presenting an output of the computerised data indicates the tidal fluctuations in the river.

1.2.4 *The unsaturated zone*

The unsaturated system

The unsaturated zone below land surface could be introduced as 'the zone where the open space in the rock is only partly filled with water'. Figure 1.17 shows the major subdivisions in this zone. Directly below land surface one finds the root zone which is usually characterised by the largest fluctuations in water content. This zone containing the network of plant roots varies in extent, but is usually less than 2 m thick. Below the root zone an intermediate zone may be present to bridge the gap with the capillary fringe. This intermediate zone can be absent in humid areas, but its thickness can also be large, especially in arid areas.

Figure 1.15. Traditional mechanical recorder showing registration chart. The recorder is installed in Lobith, The Netherlands.

The capillary fringe is the zone directly above the groundwater table. The groundwater table itself may be defined as the 'level where the groundwater pressure is equal to atmospheric pressure' (see also section 2.1.2). In the capillary fringe, as well as in the root and intermediate zone, the water pressure is less than the local atmospheric pressure. This water pressure is also referred to as 'tension' (Fetter, 1994).

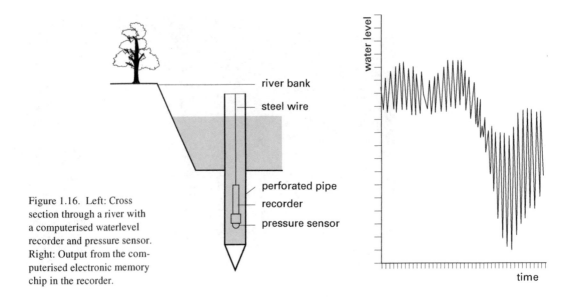

Figure 1.16. Left: Cross section through a river with a computerised waterlevel recorder and pressure sensor. Right: Output from the computerised electronic memory chip in the recorder.

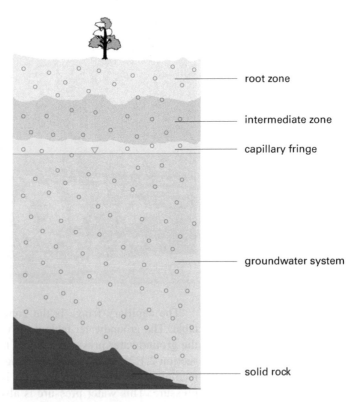

Figure 1.17. Subdivisions below land surface in the unsaturated and saturated zones.

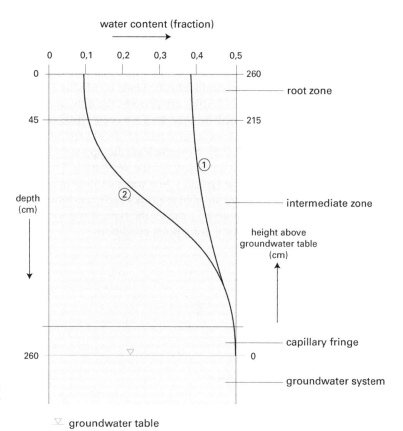

Figure 1.18. Water content distributions in the unsaturated zone for (nearly) field capacity (curve 1) and wilting point conditions (curve 2).

In the upper part of the fringe, the rock contains many pockets with air and is not saturated, but the bottom part of the fringe is saturated. The thickness of the capillary fringe usually ranges from 0.01 to 0.5 m and depends on the type of rock. In fine-grained rock material the capillary fringe is considerably thicker than in coarse-grained material. Below the capillary fringe, and thus below the groundwater table, the (saturated) groundwater system is present (see section 1.2.5).

Processes in the unsaturated zone

Part of the precipitation falling on land surface may infiltrate into the root zone and increase the water content of this zone. This water may eventually fall prone to direct evaporation or transpiration, but surplus water may also flow downward through the intermediate zone to the capillary fringe and to the groundwater table. This excess water is also referred to as 'recharge from precipitation' (see also section 1.2.5).

The condition of downward flow continues after precipitation until the so called 'field capacity' is reached. At field capacity the gravity forces acting on the water equal the surface tensions exerted by the grains and downward flow terminates. Field capacity relates to typical water contents in the unsaturated zone. Figure 1.18 (curve 1) shows the water content distribution, slightly above field capacity, for an unsaturated zone made up of silty loam.

During dry periods, the downward flow of water through the intermediate zone does not occur. Instead, there may be an upward transport of water, also referred to as capillary flow (see also sections 1.2.2 and 1.2.5). Nevertheless, the upward flow of water may not provide sufficient water for the vegetation. The so called 'wilting point' refers to the typically low water content in the unsaturated zone whereby roots are not able to extract sufficient moisture for plant survival. Figure 1.18 (curve 2) shows the typical water content distribution for silty loam, for wilting point conditions.

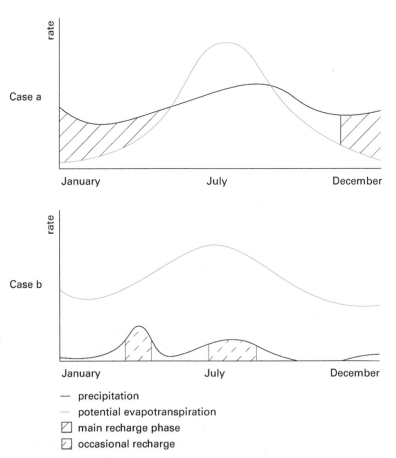

Figure 1.19. Seasonal variation in recharge to the groundwater table.
Case a: Temperate climate with main recharge in the winter season (December – February).
Case b: Arid climate with occasional recharge in the rainy season (April – August).

The water content of the unsaturated zone varies during the year. These changing conditions are related to seasonal variations in precipitation and evapotranspiration rates. Figure 1.19 (Case a) presents the conditions prevailing in temperate zones above the equator. The diagram shows that in the cold winter when precipitation exceeds potential evapotranspiration recharge from precipitation will be stimulated. In the warm summer season potential evapotranspiration exceeds precipitation. In this season, losses from the groundwater system may take place in the form of capillary flow. Figure 1.19 (Case b) presents the situation in arid areas. During the whole year the potential evapotranspiration exceeds precipitation. However, in the wet season some precipitation 'escapes' evapotranspiration, causing occasional recharge into the groundwater system.

Securing field data

Which are the parameters often measured in the unsaturated zone? These are infiltration capacity, water pressure and water content. The infiltration capacity can be measured in the field using equipment such as an infiltrometer. This instrument usually consists of a set of steel rings of 0.2 to 0.4 m diameter, which are implanted in the upper part of the unsaturated zone. Water poured into the ring will enter into this zone and 'the amount entered' is a measure for the infiltration capacity. The water pressure can be measured with a porous cup that is placed at a selected location in the unsaturated zone. A tube connects the cup to a manometer placed above ground surface where the pressure can be read. Figure 1.20 presents two types of water pressure gauging equipment. The water content, which is related to water pressure, can be determined by several methods including gravimetric, neutron, or gamma ray attenuation methods.

Figure 1.20. Tensiometers with porous cups and manometers to measure water pressure.

1.2.5 *Groundwater systems*

The physical environment

A groundwater system could be defined as 'the zone in the earth crust where the open space in the rock is completely filled with groundwater under a pressure larger than atmospheric pressure'. Such a saturated system may occupy a land surface area of only a few square kilometres. More common are the larger systems that cover areas of hundreds and even thousands of square kilometres.

A groundwater system usually 'stretches out' below the groundwater table. This table can be located near or even at land surface, but it can also be at considerable depth below the surface. One should also realise that the groundwater table is not fixed and usually fluctuates seasonally. The groundwater system extends in depth to the transition from water-containing rocks into dry and solid rocks. In areas where the transition to solid rock is at shallow depth, the thickness of the groundwater system will be rather small. In other areas the transition may be at depths of several hundreds of metres below the groundwater table forming a thick and extensive groundwater system.

Consider an interesting example of a groundwater system in the semi-arid Rada Basin in Yemen. Figure 1.21, presenting a cross section through the basin, shows that the groundwater system consists of alluvials, volcanics and sandstones (Ilaco, 1990). The groundwater system stretches out below the groundwater table, which in places is more than 50 m below land surface. In particular the locally very permeable Quaternary Volcanics offers an explanation for the deep groundwater table. The groundwater system 'ends' at the transition from the water containing sandstones into the solid and impermeable gneisses and schists. The section shows that this transition at the village called 'Noffan' is at about 600 m below ground surface. It can be deduced that the groundwater system at this village is some 500 m thick.

Groundwater recharge

Groundwater recharge can generally be considered as 'the inflow of water into the groundwater system at or near the groundwater table'. Recharge can take place in several different forms. One form of recharge, which was introduced in section 1.2.4, is recharge from precipitation caused by surplus precipitation. Recharge rates depend on precipitation and actual evapotranspiration rates, and on surface runoff. Recharge from precipitation can severely affect the groundwater table. The rise of the groundwater table may be relatively prompt or delayed, depending on the depth of the groundwater table and the rock composition of the unsaturated zone.

The effect of recharge from precipitation on the groundwater table is illustrated for a groundwater system underlying a 'sand river' in a metamorphic area in the Cape Province in South Africa (Ministry of

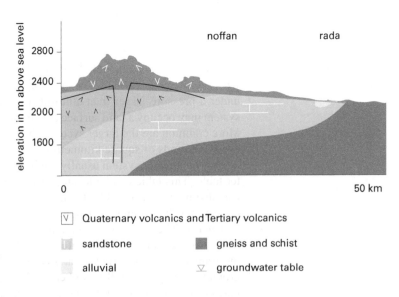

Figure 1.21. Cross section showing the geology and groundwater table in the Rada Basin, Yemen.

Figure 1.22. Pictural view of 'sand river' deposits against a background of metamorphic quartzitic rock in the Kenhardt area, South Africa.

Water Affairs, 1979). A pictural view is presented in Figure 1.22. A graph of the groundwater table measured in the period 1975 to 1977 in an observation well was prepared and is shown in Figure 1.23. The graph shows the response of the groundwater table to the seasonal precipitation and subsequent recharge. As a result of recharge, groundwater table rises in the order of 0.1 to 2 m were experienced.

Another form of recharge originates from water losses and leaching practices in irrigation schemes. Losses from irrigation include water seeping down to the groundwater table due to leakage from unlined or cracked canals or due to inefficient irrigation at the cropped land itself. Leaching is the surplus amount of water applied to the land in order to maintain a proper salt balance in the unsaturated zone. In this process extra salts are leached out which would otherwise have accumulated in this zone. Although recharge due to irrigation practices is confined to irrigated areas, its contribution to the recharge of a local groundwater system may be considerable.

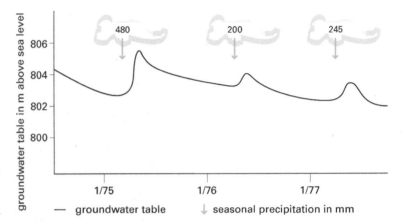

Figure 1.23. Groundwater table fluctuations as a response to recharge from precipitation (rainfall) in the 'sandy river' area in Kenhardt, South Africa.

Inflow from surface water into a groundwater system is also a form of recharge. This inflow is a result of the water level in a stream or river being at a higher elevation than the groundwater table. These conditions are also called influent conditions. Inflow rates are determined by parameters such as the difference in surface water level and the groundwater table, the wetted surface area of the stream or river bottom, and the properties of the rock adjoining the surface water system.

Seasonal variations in surface water level will affect the inflow from surface water. The inflow conditions may even be reversed to groundwater outflow conditions (see below: groundwater discharge). Attention is also drawn to the position of the surface water system in relation to the groundwater system. The surface water and the groundwater system may be in direct contact with each other, or they may exist isolated from each other. In the latter case, the groundwater table is at a much lower level than the surface water system, preventing full hydraulic contact.

The last form of recharge that will be discussed is artificial recharge. Artificial recharge can be defined as 'the practice of increasing by artificial means the amount of water that is recharging a groundwater system'. Probably the most widely used schemes for artificial recharge include infiltration ponds, and ditch and gallery systems that are filled with water from streams and rivers. This water recharges the groundwater system. After the water has travelled through the groundwater system, it is pumped to the surface using abstraction wells.

In Figure 1.24, a schematic design is presented of an artificial recharge scheme proposed for implementation in the eastern part of The Netherlands. The area is underlain by a groundwater system mainly

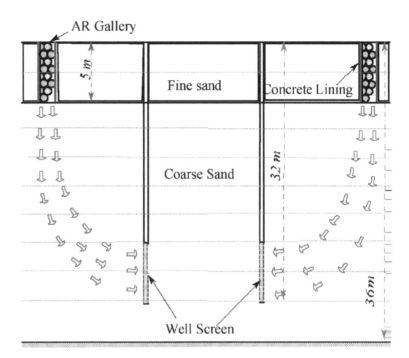

Figure 1.24. Computer print of a schematic design of an artificial recharge scheme based on galleries and pumping wells in the Vorden area, The Netherlands (Vidan-aarachchi et al., 1998).

consisting of coarse sandy material. A specially designed network of galleries takes care of recharging this system with pre-treated water obtained from a local stream. After passage through the sandy material, the recharge water will be pumped to the surface for final treatment and distribution in the region.

Groundwater discharge   Groundwater discharge can generally be defined as 'the outflow of water from a groundwater system at or near the groundwater table'. First of all, consider the form of groundwater discharge that has been utilised by old and present-day civilisations and has also appealed to the imagination of mankind: spring discharge. Although mysterious to outsiders, hydrogeologists are usually able to explain the location of a spring by picturing it within the local geological framework.

Figure 1.25 presents several geological cross sections that explain the location of a spring. Figure 1.25a shows that at the point where a groundwater table intersects the land surface a spring is formed. Figure 1.25b shows that impermeable solid basalt forces the groundwater to discharge as a spring. Figures 1.25c and d indicate the role that a fault zone can play. In consolidated rock (c), the fault may act as a permeable conduit and transport groundwater from permeable limestone to a spring zone at land surface. In unconsolidated rock (d), on the contrary, the fault and dislocated silts and clays act as an

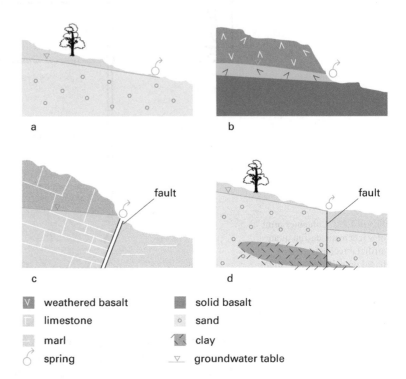

Figure 1.25. Cross sections through geological environments of a different nature, showing the most probable locations of springs.

weathered basalt		solid basalt	
limestone		sand	
marl		clay	
spring		groundwater table	

impermeable barrier forcing the groundwater to the surface, so that a spring emerges.

Springs may be classified according to the type of groundwater system they are associated with, or the magnitude of their discharge. For example, springs may emerge from groundwater systems composed of limestones, basalts or gravels (see also section 4.3). The magnitude of a spring depends on the size of, and the amount of recharge into the groundwater system contributing to spring discharge. Spring discharges increase with an increase in the size of the contributing area and with an increase in recharge.

Meinzer (1923) has suggested a classification with regard to the magnitude of springs. Table 1.4 showing this classification indicates that the majority of spring discharges are in the range from 0.75 to 250,000 m^3/day. Figure 1.26 pictures a fifth-order spring emerging from a small groundwater system in Belgium. The system is made up of karstic limestones and the small spring is one of many springs emerging from similar rock types in the area.

Although comparing magnitudes of discharges of different springs with each other is interesting, it is also quite meaningful to consider the

Table 1.4. Meinzer's classification of spring discharges.

Order	Discharge (m^3/day)		
First order	Larger	than	250,000
Second order	25,000	to	250,000
Third order	2,500	to	25,000
Fourth order	500	to	2,500
Fifth order	50	to	500
Sixth order	5	to	50
Seventh order	0.75	to	5

Figure 1.26. Fifth-order spring emerging from limestones in the Ardennes, Belgium.

discharge of a single spring in time. There are seasonal variations in discharge that may be very small, usually at the larger springs, or quite substantial. Many small springs are known to disappear in a prolonged dry season.

Another form of groundwater discharge, introduced in section 1.2.2, is the capillary flow of water from shallow groundwater tables. In the root zone this water evaporates directly from the soil skeleton or transpires through the leaves of the plants. The rate of capillary flow depends on the potential evapotranspiration rate, the depth of the groundwater table below the root zone and the type of rock material in the unsaturated zone. Perhaps the most crucial parameter in this concept is the depth to the groundwater table. For most rock types, the capillary flow from tables deeper than 1 m below the root zone is almost non-existent. Exceptions are loams where the capillary flow may be in the order of 2 mm/day from water tables as deep as 1.5 to 3 m below the root zone (Doorenbos & Pruitt, 1977). Summarising one could state that the groundwater discharge in the form of capillary flow is usually a local and temporary phenomenon. Local because the process only takes place in areas with shallow groundwater and temporary because it occurs only during dry periods when there is no recharge as a result of precipitation.

A third form of groundwater discharge is the outflow of water from a groundwater system into surface water. For the outflow into streams or rivers, this form of discharge, and any contributing spring discharge, corresponds with the baseflow that can be distinguished on a hydrograph (see section 1.2.3). Figure 1.27 shows that the outflow is a result of a surface water level being at a lower elevation than the groundwater table. This condition is also referred to as an effluent condition. The phenomenon is the reverse of surface water inflow into a groundwater system (see above: groundwater recharge). Similar to the rate of inflow into a groundwater system, the rate of outflow also depends on the difference in surface water level and the groundwater table, the wetted surface area of the stream or river bottom, and the properties of the surrounding rocks.

The last form of groundwater discharge that will be discussed is groundwater abstraction. The abstraction of groundwater can be defined by 'the efforts of men to tap the most permeable parts of the groundwater system using artificial means'. By constructing wells,

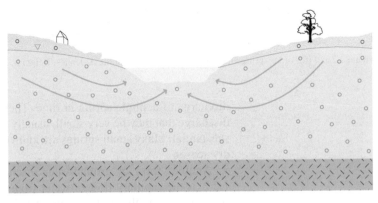

Figure 1.27. Groundwater outflow towards a river in a sandy groundwater system.

| ∘ | sand | ⟍ | clay |
| | surface water | ▽ | groundwater table |

Figure 1.28. Constructing a large diameter well by hoisting loosened material from the excavated hole. Sulawesi, Indonesia.

well fields or galleries mankind tries to secure a supply of good quality water for human, industrial or agricultural consumption.

Consider, for example, groundwater abstraction using a large diameter production well. Figure 1.28 presents a picture of the construction of such a type of well for which a hole has been excavated in permeable sandy rock material in Indonesia. The groundwater abstraction will be effectuated by lifting the water from the groundwater system using a bucket. Even a pump may be installed to bring the water up to land surface where it may be distributed.

Groundwater abstraction rates at wells, well fields or galleries vary considerably from place to place. Perhaps the most decisive factors that determine the abstraction rate are the permeability and thickness of the groundwater system. The other factors are the extent of the groundwater system, the recharge into this system, and to a certain extent the dimension of the well. In section 6.3 these factors are extensively discussed as part of the introduction to groundwater development.

CHAPTER 2

Rocks and Groundwater

2.1 BASIC CONCEPTS

2.1.1 *Physical properties of water and rock*

Water density and viscosity

Density can be defined as 'the mass per unit volume of a substance'. In the metric system the unit kg/m^3 is commonly used to express density. The density of water in nature is not constant and varies with the temperature, with the total dissolved solids (TDS) concentration, and to a very small extent with pressure. In Table 2.1, the effect of temperature on the density of water is shown, assuming a TDS concentration of 0 mg/l. The table indicates that at a temperature of 4°C, water has its highest density of 1000 kg/m^3. Above 4°C, the density decreases slightly with an increase in temperature.

In groundwater, density differences as a result of variations in the TDS concentration are more pronounced than variations resulting from changes in temperature. Table 2.2 presents computed groundwater densities for a number of TDS concentrations. It has been assumed that the dissolved solids consist of 'kitchen salt' (NaCl). In many groundwater systems this type of salt can be considered as the most representative dissolved solid. Table 2.2 indicates the increase in density for higher TDS concentrations. Densities in the order of 1000 to 1001 kg/m^3 can be associated with fresh groundwater (see also section 5.1.2) and densities in the range of 1025 to 1028 kg/m^3 are representative for saline

Table 2.1. Water density and temperature (Verruijt, 1970).

Temperature (°C)	Density (kg/m^3)
0	999.87
4	1000.00
5	999.99
10	999.73
15	999.13
20	998.23

Table 2.2. Groundwater density and TDS concentration (Oude Essink, 2001).

TDS Concentration (mg/l)	Density at 4°C (kg/m^3)
0	1000
1000	1000.7
5000	1003.6
10,000	1007.2
34,500	1025
100,000	1072

groundwater (equivalent to seawater). Densities over 1070 kg/m^3 correlate with extremely saline groundwater (equivalent to brines).

Density variations in groundwater are associated with coastal areas in many regions of the world. Figure 2.1 shows a nice example of the differences in density in the groundwater contained in the limestones of the Pedernales area in the Dominican Republic (Euroconsult, 1986). The cross section through the limestones shows that fresh groundwater with a low density is 'floating' on top of saline groundwater with a much higher density.

Viscosity can be considered as 'the property that allows fluids to resist relative movement and shear deformation during flow'. One distinguishes between dynamic viscosity and kinematic viscosity. The dynamic viscosity is usually expressed in the metric system in

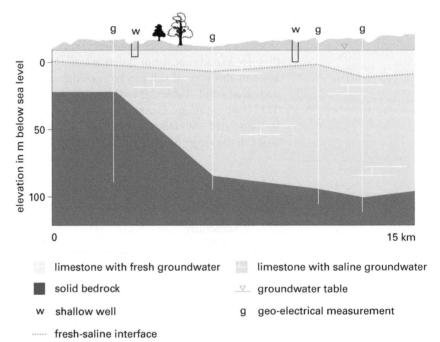

Figure 2.1. Cross section through the limestones in the Pedernales area, Dominican Republic.

limestone with fresh groundwater   limestone with saline groundwater

solid bedrock   ▽ groundwater table

w shallow well   g geo-electrical measurement

······ fresh-saline interface

Table 2.3. Viscosity and temperature (Verruijt, 1970).

Temperature (°C)	Dynamic viscosity (kg/(m*day))	Kinematic viscosity (m²/day)
0	154.66	0.155
5	131.33	0.131
10	113.18	0.113
15	98.50	0.0986
20	87.26	0.0874

kg/(m*day). The kinematic viscosity is the dynamic viscosity divided by the fluid density and can therefore be expressed in m^2/day. Table 2.3 shows the values of the dynamic and kinematic viscosity of fresh water as a function of temperature. Not surprisingly, one can deduce from Table 2.3 that the viscosity of water decreases considerably with an increase in temperature.

The structure of rocks   Groundwater is contained in rocks. These rocks are composed of minerals which all have their own particular physical appearance and chemical composition (see section 5.1.1). The most important physical features of rocks are hardness, shape and the size of the individual minerals and the way in which these minerals are arranged in the rock framework. If one considers the rock as a whole than its appearance may be compact and hard, loose and brittle, consisting of individual grains and crystals, or the rock may even have an 'amorphous' appearance.

Rocks in nature may be classified as 'consolidated or unconsolidated rock types'. The consolidated rocks are also referred to as 'hard rocks'. They are composed of solid material where the individual minerals are stuck together and can not easily be separated from each other. The unconsolidated rock types, or 'soft rocks' are made up of loose material, usually consisting of separated minerals.

Table 2.4 presents an overview of consolidated and unconsolidated rock types, with respect to origin. The table also gives examples of consolidated rocks including granite, basalt, gneiss, sandstone, shale or limestone. Unconsolidated rocks such as sands, silts, clays and loams are also taken up in the table. The consolidated or unconsolidated nature of a rock is one of the factors playing a decisive role in the determination of its groundwater characteristics.

Rock porosity   Rock porosity refers to the open space in the various rock types. In consolidated rock, openings are primarily present at fractures including joints and faults, along bedding planes, and in the form of solution holes. An illustration of selected openings in a consolidated sedimentary rock is presented at the right hand side of Figure 2.2. It is shown that the

Table 2.4. Major categories of rock types and examples.

Consolidated rock		Unconsolidated rock	
Origin	*Example*	*Origin*	*Example*
Volcanic *	Basalt		
	Rhyolite		
Intrusive *	Granite		
	Gabbro		
Metamorphic	Gneiss		
	Schist		
Sedimentary	Conglomerate	Sedimentary	Gravel
	Sandstone		Sand
	Shale		Silt
	Limestone		Clay

* Both are igneous rocks.

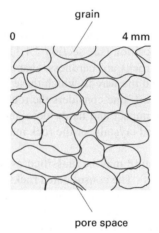 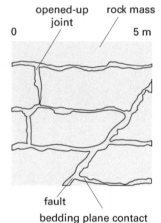

Figure 2.2. Open space in unconsolidated rock (left) and consolidated rock (right).

openings do not have an even distribution, but that they are rather localised phenomena. This type of open space in consolidated rock is commonly referred to as 'secondary porosity'.

In unconsolidated rock, openings are mainly present in between individual minerals or grains. These openings are also referred to as the pore space of the rock. The distribution of the openings is far more even than in consolidated rock. The openings can be studied at an undisturbed rock sample, using a magnifying glass or a microscope. An example of a view on unconsolidated rock material through a magnifying glass is shown at the left hand side of Figure 2.2. This type of original open space in unconsolidated rock is called 'primary porosity', which is also referred to as intergranular porosity.

The open space in rocks has formed as a result of various processes. For example, in consolidated sedimentary rock, the secondary porosity relates to the processes acting on the rock mass in the course of time. After the deposition and consolidation of loose unconsolidated sediments with primary porosity, compaction, tectonic activity and dissolution processes cause the formation of a consolidated rock mass with a secondary porosity (see section 2.2).

The porosity of a rock can also be defined as 'the ratio of the volume of open space in the rock and the total volume of rock (including the open space)'. The open space may be filled with air, water or a mix of these substances. The relationship can be expressed as:

$$n = \frac{V_O}{V_T} \tag{2.1}$$

where:
n = rock porosity (dimensionless)
V_O = volume of the open space (m^3)
V_T = total volume of rock including the open space (m^3)

In consolidated rock the porosity depends on the dimensions of the open spaces at joints, faults, bedding plane contacts, and solution openings. In addition, the amount of these openings in the rock mass is a determining factor. Table 2.5 indicates that in highly fractured and in karstic consolidated rock, the porosity may be as high as 0.5. However, in most of the consolidated rocks, the porosity is less than 0.1. In particular in dense unfractured and unjointed crystalline rock, porosities are very low (see also section 2.2.1).

Table 2.5. Porosities of common rock types.

Rock type	Range of porosities
Unconsolidated rocks	
Gravel	0.2 – 0.4
Sand	0.2 – 0.5
Silt	0.3 – 0.5
Clay	0.3 – 0.7
Consolidated rocks	
Fractured basalt	0.05 – 0.5
Karstic limestone	0.05 – 0.5
Sandstone	0.05 – 0.3
Limestone, dolomite	0 – 0.2
Shale	0 – 0.1
Fractured crystalline rock	0 – 0.1
Dense crystalline rock	0 – 0.01

In unconsolidated rocks, the porosity relates to the packing, the sorting and the shape of the grains. Figure 2.3 shows that in a 'cubic' packing arrangement, the porosity of well-sorted and well-rounded grains is in the order of 0.48. In a 'rhombohedral' framework with a similar sorting and roundness of the grains, the porosity is around 0.26. Sorting refers to the distribution of the grain sizes in a rock. The occurrence of many small grains in the open spaces between the larger grains (poor sorting) reduces the porosity. Despite differences in packing and sorting, the porosity values for unconsolidated rock generally vary within a reasonably narrow range from 0.2 to 0.7 (see Table 2.5 and section 2.2.4).

The permeability of rock and water

The permeability of a rock can be defined as 'its capacity to transmit groundwater'. The permeability of a rock depends on the properties of the rock itself and the water contained in the rock. One may first consider the part of the permeability relating to the rock characteristics. In consolidated rock, it is not only the presence of the open spaces that determines whether the rock, as a whole, is permeable. The inter-connection between the openings also plays an important role. If opened-up joints, faults, open bedding plane contacts, and solution holes, are not or hardly connected, then the overall consolidated rock can hardly be called a permeable rock.

Unconsolidated rocks are permeable where the pore space between the individual grains is sufficiently large. Large openings are normally associated with the larger grains in coarse-grained unconsolidated rocks. The openings are usually connected which implies that

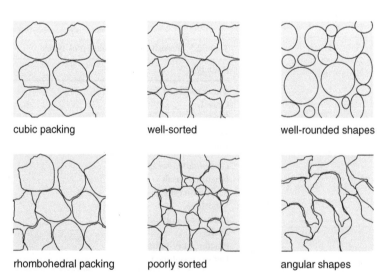

cubic packing   well-sorted   well-rounded shapes

Figure 2.3. Packing, sorting and shape of the grains in unconsolidated rock.

rhombohedral packing   poorly sorted   angular shapes

the overall rock is also permeable. Rocks with smaller openings are less permeable, also as a result of the usually plate-like appearance of their grains (e.g. in clays). The large surface area of the plate like grains 'fixes' the water molecules, which makes the rock less permeable.

The portion of the permeability relating to the water contained in the open spaces depends on the properties discussed at the start of this chapter. These properties are the water density and viscosity. The permeability is enhanced when the water has a larger density or a lower viscosity. However, variations in permeability related to changes in the properties of water are usually small.

The part of the permeability related to the rock characteristics is also referred to as the 'intrinsic permeability' of a rock. The part of the permeability related to the water can be called the 'water permeability'. The so-called 'coefficient of permeability' describes the relationship between the various components:

$$K = k\,k_{w} = \frac{k\,\rho\,g}{\mu} \qquad\qquad (2.2)$$

where:
K = coefficient of permeability (m/day)
k = intrinsic permeability (m^2)
k_{w} = water permeability (l/(m*day))
ρ = groundwater density (kg/m^3)
g = acceleration of gravity (m/day^2)
μ = dynamic viscosity (kg/(m*day))

In consolidated rock, the coefficient of permeability depends to a large extent on the widths of the open spaces at joints, faults, bedding plane contacts, and solution holes. The degree of inter-connection of these open spaces is also a determining factor. Table 2.6 shows that the coefficients of permeability of common consolidated rocks vary between 10^{-9} to 10^3 m/day. High coefficients of permeability in consolidated rocks, up to 10^3 m/day locally, are usually associated with karstic limestones and vesicular basalts. Low to medium coefficients of permeability occur in fractured sandstones and, in places, in intrusive and metamorphic rock (see also sections 2.2.1 to 2.2.4).

In unconsolidated rock, the coefficients of permeability are highly related to the grain size of the material. The coarser the material is, the higher is the coefficient of permeability. For common unconsolidated rock types the coefficients of permeability are also illustrated in Table 2.6: from 10^{-10} to 5000 m/day. It is shown that the sands and gravels have coefficients of permeability ranging from 0.1 to 5000 m/day. Fine-grained rocks including silts and clays have very low coefficients

Table 2.6. Coefficients of permeability of common rock types.

Rock type	Coefficients of permeability (m/day)
Unconsolidated rock	
Gravel	200 – over 1000
Coarse sand[1]	50 – 200
Medium to coarse sand[1]	20 – 70
Fine to medium sand[1]	5 – 30
Fine sand[2]	10^{-1} – 10
Silt	10^{-3} – 10^{-1}
Sandy clay	10^{-3} – 1
"Loose" clay	10^{-7} – 10^{-3}
Compact clay	10^{-10} – 10^{-5}
Consolidated rock	
Vesicular basalt	10^{-3} – 10^{3} [3]
Karstic limestone	10^{-1} – 10^{3} [3]
Fractured sandstone	10^{-3} – 1
Limestone, non-karstic	10^{-6} – 10^{-1}
Shale	10^{-8} – 10^{-4}
Fractured intrusive/metamorphic rock	10^{-3} – 10
Hardly fractured intrusive/metamorphic rock	10^{-9} – 10^{-5}

[1] With very low silt content. [2] With varying silt content. [3] Application of Darcy's Law equation (section 3.1) questionable for these very high permeabilities.

of permeability. They are usually less than 0.1 m/day (see also section 2.2.4). In unconsolidated material, the variation in the coefficients of permeability in the individual rocks is far less pronounced than in consolidated rocks where the coefficients can vary tremendously over short distances.

2.1.2 *Groundwater system terminology*

Aquifers and aquitards   One can classify the various rock types on the basis of their permeabilities. Such a hydrogeological classification of the rocks uses the following traditional terminology (Fetter, 2000):
– *Aquifer:* an aquifer is a rock type with a relatively large permeability. The rock is able to transmit substantial quantities of water.
– *Aquitard:* an aquitard is a rock type with a low permeability. The rock only permits the transport of groundwater in small quantities that may, however, be significant when the flow in a groundwater system is considered.
– *Aquiclude:* an aquiclude is a rock type with a very low permeability. The rock hardly transmits any groundwater although it may well contain large quantities of groundwater.

– *Aquifuge:* an aquifuge is a rock type with a negligible permeability and porosity. An aquifuge does not transmit any water; neither does it contain groundwater.

Groundwater systems normally consist of combinations of aquifers, aquitards and aquicludes. Table 2.6 indicates which rock types can be classified as aquifers, aquitards, aquicludes, or as aquifuges. For example, in many places, rocks including karstic limestones, vesicular basalts or fractured sandstones are considered as aquifers. Gravels and sands are also classified as aquifers. Fractured shales, silts, and clays are typical aquitards. If their permeability is very low, then shales and clays will be classified as aquicludes. Dense intrusive and metamorphic rock without fractures, and non-karstic and unfractured limestones can be categorised as aquifuges.

However, the classification outlined above should not be followed too rigidly. Rock types may be assigned a different classification depending on the groundwater system under consideration. For example, take the case of fine silty sand. In systems where these sediments are intercalated with clays or silts they are considered as aquifers. In other groundwater systems where similar fine silty sands are sandwiched between gravels and coarse sands, they are not considered as aquifers, but rather they are looked upon as aquitards. In addition, rocks are usually not classified in case they are located in the unsaturated zone above a groundwater system. Careful study of the geology and the permeabilities in a groundwater system leads to a correct hydrogeological classification.

To get acquainted with the concept of hydrogeological classification, two illustrative cases will be considered. Figure 2.4 shows part of the Hydrogeologic Map of China (IHEG, 1988), covering a vast area in northeast China. The area is underlain by consolidated and unconsolidated rocks. Consolidated rocks including the shown karstic and fractured limestones can be classified, in many places, as aquifers. Other consolidated rocks in northeast China composed of metamorphic gneisses and schists are impermeable in most places and can be considered as aquifuges. However, locally, these rocks are permeable due to weathering, and the presence of fractures. These localised permeable zones can be classified as local aquifers. Unconsolidated rocks in the area consist mainly of sediments deposited by the Yellow River in the North China Plain. They consist of gravels, sands, silts and clays. The saturated and non-cemented gravels and sands in these sediments can be considered as the major aquifers in the area.

The cross section in Figure 2.5 illustrates the second case. The section shows series of consolidated sedimentary rocks comprising of a succession of shales and sandstones. The upper and middle shales have a low permeability, but are nevertheless capable of transmitting small

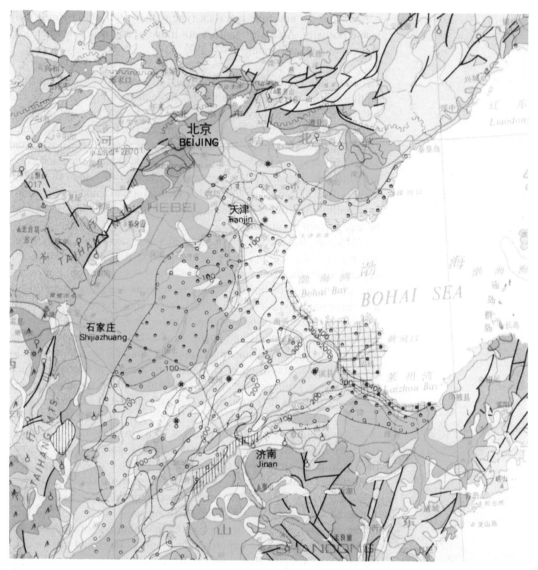

Figure 2.4. Part of the Hydrogeologic Map of China. The map shows the unconsolidated sedimentary groundwater system of the North China Plain (green colours), the fractured and karstic consolidated limestone system (blue colours) and the metamorphic rock (purple colors).

amounts of groundwater through openings at fractures including joints and faults. These upper and middle shales can be classified as aquitards. The lower shales that are less fractured have a very low permeability and can be considered as aquicludes. The fractured sandstones sandwiched between the shales are permeable. These sandstones can be considered as the aquifers of the groundwater system.

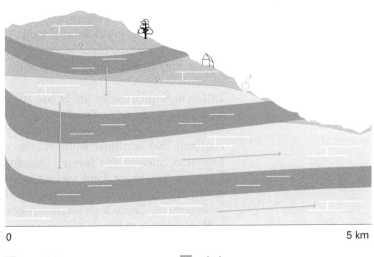

Figure 2.5. Section showing sandstone aquifers (grey unsaturated; blue saturated) and shaley aquitards and aquicludes.

sandstones

groundwater table

groundwater flow

shales

spring

Unconfined, semi-confined and fully confined aquifers

A further classification of aquifers can be made on the basis of their positions in the groundwater system. Groundwater levels measured in an observation well with its screen in the aquifer also play a role in the classification (see also section 3.1.1). The terminology is as follows:

– *Unconfined aquifers* contain groundwater that is in direct contact with the atmosphere. In most places unconfined aquifers are the uppermost aquifers. Groundwater levels measured at a well with its screen in an unconfined aquifer fall within the aquifer.

– Unconfined aquifers of an isolated nature are often referred to as *perched aquifers*. In a hydraulic sense these aquifers are not connected with other aquifers at lower levels.

– *Semi-confined aquifers* are filled with groundwater that is not directly in contact with the atmosphere. These aquifers are overlain by aquitards. Semi-confined aquifers have an inflow or outflow of groundwater through the overlying (or underlying) aquitards. The aquitards are also referred to as semi-permeable or semi-confining layers. Groundwater levels recorded at a well with its screen in a semi-confined aquifer are usually above the top of the aquifer.

– *Fully confined aquifers* are also characterised by groundwater that is not in direct contact with the atmosphere. However, these aquifers are located below aquicludes. In these aquifers there is hardly any inflow or outflow in a vertical direction. For the aquicludes the term fully confining layers is also used. When groundwater levels are registered in a well penetrating a confined aquifer, one will note that these may be well above the top of the aquifer.

Different terminologies are in use for the groundwater levels introduced above. In unconfined and perched aquifers, the groundwater levels are also referred to as 'phreatic groundwater levels'. In semi-confined and confined aquifers, the groundwater levels are also known as 'piezometric levels'. Piezometric levels in confined aquifers that rise above land surface are called 'artesian groundwater levels'. Levels of several meters above land surface are by no means an exception. These high artesian groundwater levels are associated with aquifers that have recharge areas at substantially higher elevations. The high elevations of the phreatic groundwater levels in these areas and the 'overburden load' in the lower confining parts of the aquifer cause a pressure 'built up', resulting in the high artesian levels.

The cross section shown in Figure 2.5 can also be used to illustrate the concept of aquifer classification as outlined above. Consider the upper and upper-middle sandstones in the section. Since groundwater in these aquifers is in direct contact with the atmosphere, these sandstones can be considered as unconfined aquifers. The upper sandstone aquifer is isolated from the upper-middle sandstone and can therefore also be classified as a perched aquifer. Overlain by an aquitard consisting of shales, the lower-middle sandstone aquifer can be classified as a semi-confined aquifer at the hillside where the groundwater is not in direct contact with the atmosphere and reaches the top of the aquifer. At the valley, however, the groundwater is locally in direct contact with the atmosphere. Here, the aquifer should rather be considered as an unconfined aquifer. Since the rock is overlain by an aquiclude of impermeable shales and the groundwater is not in direct contact with the atmosphere, the lowest sandstone can be classified as a fully confined aquifer.

Groundwater system boundaries

Groundwater systems may be made up of a single aquifer overlain by an aquitard or aquiclude. Alternatively, the systems may be built up of a series of aquifers separated by aquitards or aquicludes. Groundwater systems have upper and lower boundaries, and they also have boundaries in a lateral (horizontal) direction. The upper and lower boundaries of a groundwater system can, in most places, be taken as the groundwater table and the transition to a solid rock aquifuge, respectively (see also section 1.2.5). However, what are the lateral boundaries of a groundwater system? These boundaries are generally considered to be the contacts with the solid rock aquifuges that surround the groundwater system. Alternatively, lateral boundaries may be formed by the contacts with oceans and seas, lakes and rivers. The lateral transition of the groundwater system into solid rocks and open water may be rather gradual, but is in most places abrupt.

Figure 2.6 presents a case illustrating the concept of groundwater system boundaries. The figure shows a cross section across the 'sand river' in the Cape Province introduced in section 1.2.5. The section shows the small groundwater system consisting of weathered and fractured metamorphic gneiss and schist that can locally be considered an aquifer, and the alluvial sands that are aquiferous below the groundwater table. The underlying solid metamorphic gneiss and schist can be considered an aquifuge. Upper and lower boundaries of the groundwater system are respectively the groundwater table and the contact with the underlying metamorphic aquifuge. The lateral boundaries of the groundwater system are formed by the northwestern and southeastern contacts with the unweathered metamorphic aquifuge.

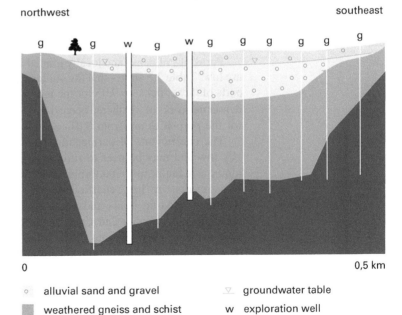

Figure 2.6. Cross section through a sand river showing alluvial, gneiss and schist, Kenhardt Area, South Africa.

o alluvial sand and gravel

weathered gneiss and schist

solid gneiss and schist

▽ groundwater table

w exploration well

g geoelectrical measurement

2.2 FORMATION OF GROUNDWATER SYSTEMS

2.2.1 *Groundwater systems in metamorphic and intrusive rock*

Groundwater in metamorphic rock

What details of groundwater as contained in consolidated metamorphic rock are noteworthy? First, take a look at geology. As a result of intense pressure and high temperatures sedimentary rock can be transformed into metamorphic rock. Scientists have also suggested that metamorphic rock can be recrystallised from igneous rock. Common metamorphic rock types include schists, gneisses, quartzites, slates, and marbles. The main mass of these rocks is made up of solid and dense material that may be intensively foliated. Figure 2.7 shows a close up of a gneissic rock that is part of a vast metamorphic rock complex in southeast India. The picture shows the dense nature of the rock mass.

Nevertheless, open spaces may be present at joints and in places where metamorphic rocks are faulted. Weathering in metamorphic rock can affect the open spaces. In particular in the fine-grained rocks such as slates and schists weathering may extend up to depths over 100 m into the fresh bedrock. Weathering in quartzites is much less pronounced or it is not present at all. The effect of weathering is that fractures are further opened-up and inter-connected and that the minerals that make up the bedrock are (partly) transformed into gravels, sands, clays and laterites. These processes tend to increase the porosity and the permeability of the rock.

Porosities of dense and locally jointed crystalline metamorphic rocks are low. Table 2.5 shows that the values range from zero for the denser parts of these rocks, to 0.05 for the more jointed zones. For rocks with denser networks of weathered opened-up joints and faults common ranges for the porosities are from 0.01 to 0.1.

The coefficients of permeability for metamorphic rocks are usually low and may be near to zero. One has to realise that opened-up joints and faults in metamorphic rock may not be connected which also results in low coefficients of permeability. Nevertheless, where opened-up joints and faults are connected, the rock may be reasonably permeable and coefficients of permeability up to 10 m/day have even been reported. Abstraction wells drilled in metamorphic rock should be sited with utmost care in order to penetrate permeable zones. Abstraction rates of wells tapping these zones could be in the order of 25 to 300 m^3/day.

Metamorphic rocks are generally considered as poor groundwater systems. The rocks mainly consist of aquifuges. However, parts of metamorphic rocks may be classified as local aquifers in places where the rock mass has a dense network of opened-up fractures, which may be affected by weathering.

Groundwater associated with intrusive rock

Details of groundwater occurrences in consolidated intrusive rock can be considered. But first, what is intrusive rock? Intrusive rocks belong to the group of igneous rocks, which are rocks of magmatic origin. In intrusive rocks the fluid magma in the earth crust has hardened before it reached land surface. This resulted in the formation of dense crystalline rock in which the individual mineral grains can usually be observed by eye. Common intrusive rock types are granite, gabbro and diorite.

Figure 2.7. Close-up of gneiss showing the dense nature of the rock, especially at the upper ballpoint pen. Hirakud area, India.

Figure 2.8 has been composed to point out the open space, present in intrusive rock. The block diagram shows that openings are present at opened-up joints, faults and a weathered zone in granitic rock. The joints may include cooling joints that have opened up as a result of pressure relief, and the faults may have formed due to intense lateral pressure. In the weathered zone the open space at these fractures has been enlarged and the rock may even have turned into loose material. Although not shown in the diagram, fracture patterns may also be present in contact zones between the intrusive rock and invaded 'country rock'. Porosities in intrusive rocks are much the same as in metamorphic rocks and range from zero for the denser parts, to 0.1 for the fractured and weathered zones.

Coefficients of permeability in intrusive rocks are also similar to those in metamorphic rocks. Values range from zero for the solid parts to 10 m/day for fractured and weathered rock masses. Acceptable groundwater abstraction rates can be secured in intrusive rocks where wells penetrate connected opened-up joints or tap from well-developed fault zones. Useful abstraction rates may also be obtained at wells drilled into the weathered zones of the intrusive rocks themselves and into permeable associated rock of a different composition. High well abstraction rates may, for example, be found in gravels or marbles associated with intrusive rock.

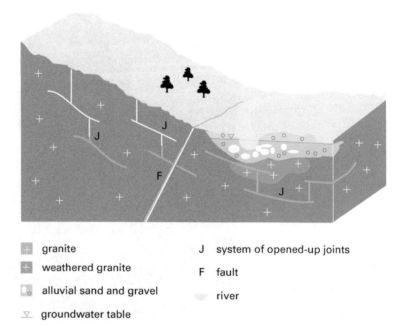

granite

weathered granite

alluvial sand and gravel

▽   groundwater table

J   system of opened-up joints

F   fault

river

Figure 2.8. Block diagram showing areas with openings in granitic rock.

Intrusive rocks are generally not classified as groundwater systems containing abundant aquifers. Similar to metamorphic rock, the 'intrusives' can generally be looked upon as aquifuges whereby at fractured or weathered zones local aquifers may have formed. The local aquifers usually occupy rather narrow zones that may extend, however, over large distances.

The relation between depth and permeability

In metamorphic and intrusive rocks the porosity and permeability tend to decrease with depth. This phenomenon can be explained by the tendency of fractures to close as a result of the weight of the overburden, and the decrease in weathering with increasing depth.

Figure 2.9 illustrates this relationship by giving consideration to abstraction rate. The right diagram indicates that the increase in abstraction rate per unit depth interval decreases with depth. The decrease in the increase in abstraction rate points to a decrease in permeability with depth. The left diagram illustrates the decrease in permeability, showing less opened-up fractures with increasing depth.

The relationship sketched above is not always valid. One may think, for example, of a permeable fractured zone encountered in a well that has 'first' been drilled through impermeable dense crystalline rock. Despite the fact that fractures tend to close with increasing depth it is well known that some open fractures exist at depths in the order of several hundreds of meters below land surface. Note that wells drilled in quartzitic rock in a metamorphic belt in South Africa produced groundwater at a rate over 400 m^3/day from a fractured zone more than 100 m deep (Ministry of Water Affairs, 1977).

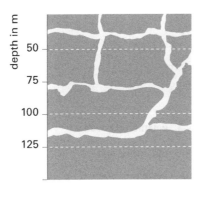
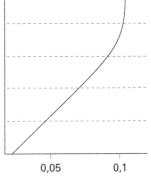

Figure 2.9. Relationship between increase in abstraction rate and depth below land surface.

rock section

increase in abstraction rate per unit depth in L/sec/m

2.2.2 *Development of aquifers in volcanic rock*

Groundwater in lava flows

What can be said about groundwater in consolidated volcanic rock? First, consider the rock itself. Igneous rocks originating from magma that has erupted at earth surface is called volcanic rock. Magma flowing over the rim of volcanoes and crystallising at land surface is called lava. In many places lava flows are manifested as a sequence of horizontal or sub-horizontal layers originating from different eruptions. A sequence may be several hundreds of meters thick. The upper parts of the individual layers may be weathered and even old soil profiles may be visible. Also, the individual layers or flows are in many places separated by alluvial deposits or other materials. A common type of volcanic rock is rhyolite, but even better known is basalt.

To be able to visualise open space in volcanic rock, one has to realise the layered structure of volcanic rock. The relation between open space and layering will be illustrated using an example. Figure 2.10 shows a block diagram through basalt indicating two basalt layers. The diagram shows that within a layer the rock may be dense, but may also have openings at places with pronounced opened-up joints

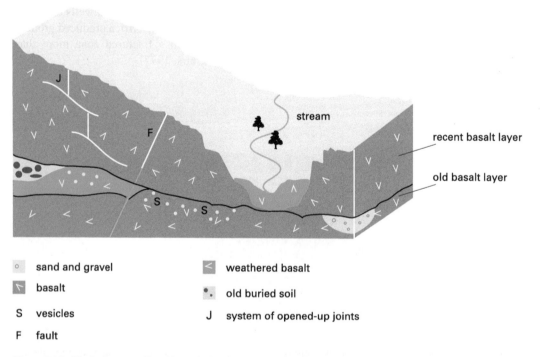

○	sand and gravel	<	weathered basalt
⌐	basalt	•.	old buried soil
S	vesicles	J	system of opened-up joints
F	fault		

Figure 2.10. Block diagram of basaltic rock showing two lava flows and areas with rock openings.

and faults, and at horizons with gas holes that are also referred to as 'vesicles'. At the contact between individual layers where weathered zones and soils have formed (old buried soils), and where sands and gravels have been deposited, the rock may also have openings.

The uneven distribution of open space implies that porosities of volcanic rocks vary considerably. In some areas, the rocks may be extremely dense and have porosities far less than 0.01. In other areas, the porosities may be relatively high. Table 2.5 shows that porosities of fractured basalt may be in the order of 0.05 to 0.1, and even up to 0.5.

Within the volcanic rock mass, the permeability may change tremendously over short distances. The more permeable parts of volcanic rocks are present in places where opened-up joints and faults, vesicles, open space at weathered parts, etc are well-connected. Especially, large connected fracture systems may result in high permeabilities. Highly permeable parts may also be present at the sandy or gravelly intercalations in between the volcanic layers. Buried clayey soils at the contact between the volcanic layers are usually not very permeable. Less permeable are also dikes (intrusive rock with the form of a dike or wall) that may have intruded into the volcanic rock. However, at the contact zone between a dike and the volcanic rock, permeable zones may be present. The large variations in permeability in volcanic rocks are also reflected in the ranges for the coefficients of permeability. Table 2.6 shows that coefficients of permeability of basalts range from 10^{-3} m/day for the denser rock types with little vesicles, to 10^3 m/day for rocks with abundant vesicles, fractures and weathering.

Occurrence of ground-water in pyroclasts

There are other volcanic rocks than the rocks that originate from lava flows. Part of the magma that erupts at volcanoes at the earth surface is thrown into the air. Clouds of volcanic material may contain volcanic bombs, ash and even smaller sized particles that are subsequently 'deposited' on the slopes of the volcanoes and the areas beyond them. These deposits are usually referred to as pyroclasts. Special forms of pyroclasts are, for example, tuff, pumice and volcanic ash beds. In particular in places where the magma is acid, pyroclasts have been deposited.

The open space in pyroclastic deposits may vary significantly. Unwelded tuffs usually consist of loose angular and unsorted fragments. The open space in these rocks may be moderate and even high. Primary porosity is usually present in these rocks. In welded tuffs the fragments are molten together and the open space is usually less than in the unwelded tuffs. However, in places, the welded tuffs have secondary porosity at opened-up fractures. Pumice and volcanic ash deposits have a high percentage of open space. Table 2.7 shows the

Table 2.7. Properties of the Oak Springs Group in Nevada (Keller, 1960).

Lithology	Density (kg/m^3)	Porosity	Coefficient of permeability (m/day)
Bedded tuffs, zeolotized	1500	0.39	$3.5 * 10^{-5}$
Bedded tuffs, pumiceous	1370	0.40	$1.0 * 10^{-2}$
Friable tuffs	1500	0.36	$1.2 * 10^{-3}$
Welded tuffs	2180	0.14	$2.8 * 10^{-4}$

relatively high porosities for the Oak Springs Volcanic Group in the United States, ranging from 0.14 to 0.40.

The pyroclastic deposits usually have a low permeability. The grain size of unwelded tuffs is rather small which is reflected in a reduced permeability. The welded tuffs may be somewhat permeable where the fractures are inter-connected and cover a larger area. Pumice and volcanic ash beds may have a substantially larger permeability. Table 2.7 shows that the coefficients of permeability of pyroclasts for the Oak Springs Volcanic Group range between $3.5*10^{-5}$ and 0.01 m/day.

Volcanic groundwater systems

In volcanic groundwater systems, one may have sequences of lava flows, interbedded with sandy or pyroclastic deposits, or there may be just thick deposits of various types of pyroclasts. The high variability in composition also implies that volcanic groundwater systems usually consist of an alteration of aquifers, aquitards and aquicludes. The lava flows contain thin aquifers which are usually the subhorizontal zones with opened-up fractures, and with vesicles. Thin aquifers are also formed by the weathered profiles, the interbedded series of sandy and gravelly material or pumice. Pyroclastic ash beds may be less permeable and act as aquitards. Tuffs interbedded in these volcanic groundwater systems may play a dual role: in most places they would act as aquitards or aquicludes. In other places, for example where they are sandwiched between aquifuges of impermeable columnar basalts, tuffs may also take on the role of an aquifer of limited capacity.

Figure 2.11 shows the variation in volcanic rock composition in the Sanaa Basin in Yemen. The picture shows layers of reasonably dense basalt on top of bedded pyroclastic deposits. The basalt is dense, but may be locally jointed, whereas the pyroclasts are unwelded and have some open spaces in between the grains. Both the basalt and the pyroclasts may locally form aquifers below the deep groundwater table in the Sanaa Basin.

The relation between geological time and permeability

The older the volcanic rocks, the less permeable they usually are. In many areas the weight of younger overlying rock, the so-called 'overburden', has caused a reduction of the permeability of the older volcanic rocks. The overburden consists of younger volcanics or sedimentary deposits. In areas where the older volcanic rocks are not overlain by younger deposits, chemical processes may have played a role. Minerals dissolved in groundwater may have precipitated in opened-up fractures, in the vesicles, and in lava tubes, which may also have resulted in a decrease in permeability.

In many areas of the world this relationship holds. In India, in the Eastern United States, and in the Rift Valley in East Africa and the Saudi Arabian Peninsula, basalts from the Tertiary age have a much lower permeability than the younger Quaternary volcanic rocks. As a consequence abstraction rates, but also spring flows in recent Quaternary volcanic rock in these areas are larger than in the older Tertiary rock. Traditionally, but also in the future, any consideration of large-scale groundwater development will focus on the exploitation of the most recent volcanic deposits.

2.2.3 *Aquifer formation in consolidated sediments*

Groundwater occurrence Groundwater in consolidated fine-grained sedimentary rock is worth
in fine-grained sediments discussing. What can be said about the origin of these rocks? In

Figure 2.11. Section through volcanic deposits in a road cut in the Sanaa Basin, Yemen.

the course of time most fine-grained unconsolidated sediments have turned into consolidated sediments. Clays have turned into claystones or shales, and silts and mud have turned into siltstones and mudstones. Shales, claystones, and mudstones account for almost 50% of all the consolidated sedimentary rock. In many areas, the layers of fine-grained sediments are alternated with coarse-grained sediments or carbonate deposits. The layers may be folded into anticlines and synclines and faulting may also be present.

When consolidated fine-grained sediments are inspected under the microscope, one will note that primary porosity is not significant. Open space in these rocks is rather found as secondary porosity at opened-up joints, at fault zones, and to some extent at contacts between the individual beds. Weathering may also increase the open space in fine-grained rocks. Table 2.5 shows that values for porosities in fine-grained shales are in the order of zero to 0.2. Table 2.8 showing the porosities for a number of shales in the United States confirms these values.

Although the porosities of consolidated fine-grained sediments are not insignificant, the coefficients of permeability are usually very low as is also shown in Tables 2.6 and 2.8. Values for the coefficients of permeability of fine-grained sediments are usually less than 10^{-4} m/day. Nevertheless, the coefficients of permeability may be slightly higher where fine-grained sediments have inter-connected fractures, or where moderate weathering has taken place. In these places, the coefficients of permeability may be large enough to supply minor amounts of water to wells. Abstraction rates in the range of 25 up to 50 m^3/day have been reported.

Coarse-grained sediments and groundwater   The occurrence of groundwater in consolidated coarse-grained sediments is widely known. How were these sediments formed? Through the action of pressure and temperature coarse sands and gravels have transformed into sandstones and conglomerates. Sandstones are widely spread consolidated sedimentary rocks, whereas conglomerates are less common. Sandstones can be composed of angular or rounded quartz grains, but other minerals including feldspars or micas may also be present.

In coarse-grained consolidated sediments, the space around the grains may be open, or cemented to a certain degree. The cement, which reduces primary porosity in these rocks, can be present in the form of calcite, silica or clay. These minerals have usually precipitated from groundwater. There are also reports from Turkey that groundwater dissolves the cement, restoring the original pore space of the rock. In addition to primary porosity, coarse-grained consolidated rocks also exhibit secondary porosity that can be observed at opened-up joints, at faults and at the contact between bedding planes. In general,

Table 2.8. Porosities and coefficients of permeability of consolidated sediments (Davis, 1967).

Rock type	Porosity	Coefficient of permeability (m/day)
Shale, Cretaceous/USA	–	$3.5*10^{-6}$
Shale, Gros Ventre/USA	0.11	–
Shale, Graneros/USA	0.25	–
Shale, Pre-Cambrian	0.02	–
Sandstone, Tawilah/Yemen	–	0.2
Sandstone, Berea/USA	0.19	0.3
Sandstone, Ordovician/USA	0.07	$3.5*10^{-3}$
Sandstone, Cretaceous/USA	0.26	3.8
Sandstone, shaley/USA	0.15	$2.6*10^{-4}$
Limestone, compact, some pores/USA	0.10	$3.5*10^{-3}$
Limestone, chalky/USA	0.30	$3.3*10^{-2}$
Limestone, oolitic	0.22	0.3

coarse-grained consolidated sediments have porosities similar to those of fine-grained rocks. Table 2.5 shows that the porosities of sandstones range from 0.05 to 0.3. Table 2.8, also presenting porosities for sandstones in the United States, shows values which are in the same range.

To a large extent, the degree of cementation determines the permeability of a coarse-grained rock. In addition, the presence of networks of opened-up fractures and open spaces at the contacts between bedding planes increases the permeability. Coefficients of permeability of coarse-grained consolidated sediments are usually one or two magnitudes larger than the coefficients of permeability for fine-grained sediments. For example, Table 2.8 shows that the coefficients of permeability of sandstone rocks in Yemen and in the United States are considerably larger than the coefficient of permeability shown for the Cretaceous shale. Tables 2.6 and 2.8 indicate that the general range for the coefficients of permeability for sandstones is in the order of 10^{-4} to 5 m/day.

Figure 2.12 shows an oval shaped basin surrounded by coarse-grained sandstones with extensive networks of connected opened-up fractures. The sandstones belong to the Tawilah Formation and are located in the central part of Yemen. Springs frequently emerge at the base of the permeable networks, and at contacts with less permeable rock. Coefficients of permeability for the Tawilah Formation are generally in the range of 0.1 to 0.5 m/day, and have been determined from pumping tests and modelling studies (NWASA & TNO, 1996).

Common abstraction rates of wells drilled in coarse-grained consolidated rocks range between 50 and 750 m³/day. However, smaller

Figure 2.12. Oval-shaped valley, cut out in aquiferous Tawilah sandstones. Small springs are indicated in the background where the village is located, Kawkaban area, Yemen.

and larger rates than indicated are by no means an exception. The safest way to drill a successful well is to select its location in non-cemented sandstone with a high primary porosity. Success rates for the drilling of wells are much smaller in sandstones where secondary porosity is dominant in the form of open space at fracture zones and at bedding plane contacts. In these types of rocks, extensive investigations are needed to locate sites for wells with adequate abstraction rates (see section 2.3).

Groundwater in carbonate rocks

The occurrence of groundwater in consolidated carbonate rocks has puzzled hydrogeologists for many years. Carbonate rocks mainly include limestones and dolomites. These rocks originate from the shells of living organisms which have been deposited on the bottom of the sea, from chalky coral reefs or from the chemical precipitation of calcium, magnesium and carbonate ions. Carbonate rocks are mainly formed in marine environments.

Open space is usually present between the calcareous shells and reef fragments during the deposition of carbonate rocks. Rocks with such open spaces have a primary porosity. After these 'soft' deposits including chalks have been consolidated into limestones and dolomites secondary porosity becomes more important. Secondary porosity in consolidated carbonate rocks can be observed at opened-up joints formed in the crest of an anticline or in the saddle of a syncline, at fractures associated with faults, and at bedding plane contacts. Figure 2.13 shows the joints in the anticline of a series of limestones exposed

Figure 2.13. An anticline in the limestones of the Ardennes, Belgium.

in the eastern part of Belgium. Secondary porosity may further develop at opened-up joints, faults and bedding plane contacts by the dissolution of carbonates (see section 5.1.1). In this way large solution holes are being formed as part of the so-called 'karstification process'. Generally, solution holes are better developed along sub-horizontal bedding plane contacts and especially along fault zones, than along vertical joints.

Figure 2.14 illustrates the karstification process showing a view across almost sub-horizontally bedded limestones located in the Jericho area near the Dead Sea in the Middle East. In places, solution holes are visible as dark caves on the picture. The solution process has been so severe that parts of complete beds have been 'eaten away'. The dissolution did take place at a time when the rock was below the groundwater table. Later uplift of the rock mass exposed the karst features to the open air (see also section 5.2).

Porosities of carbonate rocks are quite variable as a result of the large differences in open space at fractures. The degree of karstification may also play a decisive role. Carbonate rocks with limited open spaces have porosities less than 0.3. Table 2.8 indicates that the porosities of three

Figure 2.14. Karstification in the Cenomanian-Turonian limestone complex in the Eastern Basin in the Dead Sea Rift Valley, Middle East.

selected limestone series in the United States vary between 0.10 and 0.30. On the other hand, Table 2.5 indicates that porosities for karstic limestones are more variable and range from 0.05 to 0.5.

Coefficients of permeability for carbonate rocks show large variations. Table 2.6 shows the typical ranges for the coefficients of permeability in karstic limestone rocks: from 10^{-1} to 10^3 m/day. Permeabilities for the limestone rocks compiled in Table 2.8 range from $3.5*10^{-3}$ to 0.3 m/day. The coefficients of permeability for shaley limestones are usually at the low end of the range, whereas the coefficients of permeability of jointed, faulted, and karstic limestones as well as the coefficients of the young so-called 'reef' limestones are at the upper end of the series. Coefficients of permeability for other carbonate rocks may fall outside the quoted ranges. For example, experiments carried out in the United States have shown that permeabilities for dolomite rocks may be higher than the typical values for comparable limestones (Davis & De Wiest, 1967).

The coefficients of permeability listed in Tables 2.5 and 2.8 refer to average values for a whole rock series at a particular location. However, it should be realised that on a smaller scale, extremely large variations in coefficients of permeability are normal. As a result of the 'closeness' of open joints, faults, bedding plane contacts and solution holes on one hand, and dense massive rock on the other hand, the coefficients of permeability may vary from zero to over 100 m/day, within the range of a few metres.

Due to the high variability of the coefficients of permeability in these rocks, abstraction rates will also differ significantly over short distances. Dry wells can be drilled near wells yielding large amounts of water. The narrow margin between success and failure in drilling was amply demonstrated in a dolomite rock in the western part of Transvaal Province in South Africa. Wells drilled into karstic solution holes and into fractures indicated by quartz veins produced yields in the order of 1500 to 2500 m³/day, whereas wells drilled into the contact zone between dolomite and dolerite yielded only 50 to 100 m³/day. Wells drilled into the non-karstic and non-fractured parts of the rock were dry (Ministry of Water Affairs, 1977).

Aquifers and aquicludes in consolidated sediments

Most of the consolidated rocks originally formed in a sedimentary groundwater system are combinations of fine-grained sediments, coarse-grained sediments, and carbonate rocks. There is no fixed rule, but one generally assumes that fine-grained rocks in a sedimentary groundwater system stand out as aquitards or aquicludes. In many systems, the coarse-grained sediments and carbonate rocks are the aquifers. However, one should not overlook the fact that in some sedimentary groundwater systems, the coarse-grained and carbonate rocks cannot be considered as aquifers. Dense coarse-grained sandstones and shaley carbonate rocks in these systems are not classified as aquifers, but may rather be categorised as aquitards or even as aquicludes.

2.2.4 *Unconsolidated sediments and groundwater systems*

Fine-grained unconsolidated rock and groundwater

How relates groundwater to fine-grained unconsolidated deposits? Fine-grained rocks may include clays, silts, loams, and other materials with a fine texture. Depending on the classification used, these rocks have grain size diameters smaller than 0.05 mm. The sediments may have been deposited in the sea, in lakes or they may have settled out after floods in river areas.

The openings in fine-grained unconsolidated sediments are formed by primary porosity. Through compaction and stress the primary porosity of fine-grained deposits quickly decreases with time. Tables 2.5 and 2.9 indicate that clays and silts have large porosities compared to the porosities of consolidated sediments and other unconsolidated rocks. The values for porosities of silts and clays are in the order of 0.3 up to 0.7. For dense and well-packed clays, the porosities agree with the bottom end of this range: 0.3 to 0.4. In 'loose' clays where the packing of the particles is very irregular, the porosities may be as high as 0.7.

Coefficients of permeability of fine-grained sediments are usually low. Tables 2.6 and 2.9 show that clays have coefficients of permeability in the range of 10^{-10} to 1 m/day. In places where clays have been deposited without coarser material, the coefficients of permeability correlate with the lower end of the range: from 10^{-10} to 10^{-4} m/day. Fine-grained silts composed of larger grains than clays have higher coefficients of permeability. As shown in Table 2.6, common values range from 10^{-3} to 0.1 m/day. Loosely packed loess deposits of the silt fraction, however, can even have coefficients of permeability in the order of 0.2 to 0.3 m/day.

Groundwater and unconsolidated coarse-grained sediments

Groundwater occurrences in unconsolidated coarse-grained sediments are well known. These sediments which are mainly deposited on the continents and in the sea near the coast line include sands and gravels. Common grain sizes of sands are in the order of 0.05 to 2 mm diameter. In this interval one can differentiate between fine, medium and coarse sand. Gravels have grain size diameters over 2 mm.

Coarse-grained sediments have distinct open spaces between the grains. Primary porosity is dominant in these types of rocks. The porosity of coarse-grained deposits depends on packing, but in many places the sorting will play an even more important role. Poor sorting of a coarse-grained deposit may reduce the porosity by almost half. Tables 2.5 and 2.9 indicate that common ranges for the porosities of coarse-grained sediments are in the order of 0.2 to 0.5. These values are in

Table 2.9. Relation between rock type, porosity and coefficient of permeability as determined by laboratory tests on collected samples (Davis, 1967).

Sample	Origin	Rock type	Porosity	Coefficient of permeability (m/day)
1	Marine	Clay	0.48	$1.4*10^{-5}$
2	Loess	Silt	0.49	0.19
3	Loess	Silt	0.51	0.28
4	Alluvium	Fine sand	0.45	11.4
5	Alluvium	Fine sand	0.52	4.7
6	Marine	Medium sand	0.42	33
7	Alluvium	Coarse sand	0.33	164
8	Alluvial	Gravel	0.25	975

the same order, or tend to be somewhat lower than the porosities for fine-grained deposits.

Permeabilities of coarse-grained deposits, however, are large compared to those of fine-grained unconsolidated sediments and compared to most consolidated rock types. Most striking, especially in coarse-grained sediments, is the direct relationship between permeability and grain size. The coarser the grain size is, the larger is the permeability (see also section 2.1.1). Tables 2.6 and 2.9 confirm this relationship and further show that the coefficients of permeability of sands are in the range between 1 and 200 m/day. Common ranges for gravels are from 200 to over 1000 m/day.

Table 2.10 has been prepared to provide information on coefficients of permeability in the eastern part of The Netherlands (IHE, 1992–2000). The area is underlain by vast deposits of sands and gravels. The table indicates that the coefficients of permeability of the fine sandy and coarse sandy sediments are in the ranges of 5 to 10 m/day, and 40 to 60 m/day, respectively. The coefficients of permeability should be considered as regional values, since they were determined on the basis of groundwater modelling studies covering extensive areas.

In many places, wells tap unconsolidated coarse-grained sediments. These coarse-grained rocks may cover extensive areas. Sediment thicknesses of a few meters to over 100 m and more may be encountered.

Table 2.10. Relation between origin, rock type and regional permeability in Eastern Netherlands.

Area	Origin	Rock type	Regional permeability (horizontal coefficient of permeability in m/day)
Southeast Drenthe	Fluviatile alluvium	Coarse sand with some gravel	60
Southeast Drenthe	Periglacial	Fine sand with some loam lenses	10
Zwolle City Area	Fluviatile alluvium	Coarse sand/medium sand/some gravel	45
Weerselo Area	Eolian	Fine sand	4
South Veluwe	Fluviatile alluvium	Mainly coarse sand	52

Abstraction rates in wells in saturated sands and gravels normally range between 400 and 4000 m^3/day. In case sufficient recharge is available, then any sizable extent of the deposits guarantees a continuous flow of water to a well. Leakage of water from fine-grained sediments above the coarser deposits may also contribute to well abstraction. It is not surprising that coarse-grained unconsolidated sediments are favoured for the installation of wells and well fields.

Aquifers and aquitards in river valleys, tectonic basins and coastal plains

Groundwater systems consisting of unconsolidated rock are usually associated with river valleys, tectonic basins or coastal plains. Different forms of sedimentation have taken place in these areas resulting in sets of deposits which may be vastly different with respect to grain size, sorting, packing etc. In fact all sorts of deposits including continental stream, river, lake, wind-blown (eolian) and glacial deposits, and marine coastal and deep sea deposits may be present.

River valleys that are primarily formed by erosion are usually filled with unconsolidated deposits that can be of considerable thickness. Figure 2.15 shows a cross section through a river valley indicating the succession of unconsolidated material. The coarsest sediments, deposited by the river, are usually found at the bottom of the valley. Here, during the early stages of river formation the slope of the terrain was still large and the coarser sediments were deposited, while the finer material was carried further downstream. Other coarse sediments are associated with the riverbed itself or with old riverbeds that have been covered with much finer material. When inspecting the more superficial layers, then, from the river proceeding land inward, one will usually find a transition of fine to medium sands at the river levees to fine silts and clay plugs in the oxbow lake areas.

Figure 2.15 can also be considered to illustrate the classification of aquifers and aquitards, in a river basin. The coarse sands and gravels

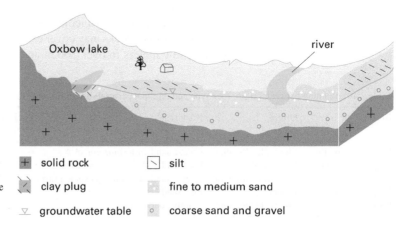

Figure 2.15. Unconsolidated sediments above impermeable bedrock in an eroded river valley.

+	solid rock	◳	silt
◿	clay plug		fine to medium sand
▽	groundwater table	◦	coarse sand and gravel

at the bottom form an aquifer and the fine silts and clays at the top represent an aquitard. Although this illustration is instructive it is not representative for the general built up in all river valleys. Especially in large river systems, the groundwater system usually consists of several sandy and gravelly aquifers separated by silty or clayey aquitards. Dense compact clays may even form nearly impermeable aquicludes.

Tectonic basins have formed as a result of the downward movement of parts of the earth crust along major fault systems. During and after these movements tectonic basins may have been filled with deposits, delivered by streams from surrounding mountains. Even volcanic rock may be part of the basin deposits. Coarse-grained stream and river deposits, sandy to gravelly foothill deposits, buried dune sands or vesicular basalt may occur, as well as fine-grained flood plain deposits, lacustrine silts and clays, or columnar basalt.

Groundwater systems contained in tectonic basins are complex and usually consist of an interfingering of aquifers, aquitards and aquicludes. The coarse-grained deposits can be classified as aquifers, whereas the fine-grained deposits will form aquitards or even aquicludes. Fine-grained silts, loams and clays in tectonic basins may occur only locally, as lenses, thereby reducing the extent of aquitards and aquicludes. In tectonic basins, the groundwater system may consist of a single aquifer, but in most places several aquifers, aquitards and aquicludes are present.

Coastal plains may contain successions of continental and marine sediments. Continental deposits may include beach sands, shore line sediments, and stream and river deposits from mountain ranges bordering the coastal zone. These continental coastal deposits are usually coarse-grained and may grade from gravel to fine sand. The marine sediments may be made up of deltaic, lagoonal or deep-sea deposits consisting of finer graded materials like silty sands, silts and clays. Carbonate coral reef and chalky rock may also be present. In the coastal deposits an interchanging with other rock types including volcanic types is also common.

Groundwater systems contained in coastal plains are usually complex systems composed of several aquifers, aquitards and aquicludes. The complexity of coastal systems is a result of the large variety in coastal sediments that can be deposited. It is not only the variety of sediments, but it is also their irregular occurrence that makes assessments with regard to groundwater systems in coastal areas difficult. A further complication in coastal groundwater systems is that they contain brackish or saline groundwater.

Figure 2.16 presents an example of a typical groundwater system in a coastal area. The figure shows a cross section through the coastal dune area in The Netherlands. The 'Calais' and 'Strandvoet' Formations are fine sandy and locally silty lagoonal and beach sediments that

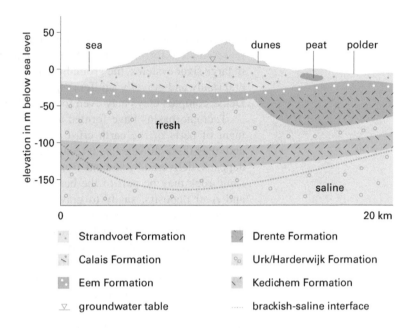

Figure 2.16. Cross section through the coastal area in The Netherlands (Geochem, 1992).

form the shallow aquifer in the area. The 'Eem' and the 'Drente' Formations are respectively of marine and glacial origin, partly consisting of silty and clayey aquitards and aquicludes. The 'Urk' and 'Harderwijk' Formations form the bottom aquifer mainly consisting of medium to coarse sandy stream and river deposits. It is interesting to note that the present day interface between brackish and saline water is located in the 'Harderwijk' Formation. The saline water in these sediments is presumably 'exchange' water that invaded into the groundwater system when the marine 'Eem' and 'Calais' Formations were deposited (see also section 5.1.2).

2.3 GROUNDWATER EXPLORATION METHODS

2.3.1 *Methods at land surface*

Desk studies and preliminary field reconnaissance

The occurrence of groundwater can be investigated by desk studies and a number of field methods carried out at land surface. Desk studies are undertaken in the office on the basis of available information. Groundwater related information is contained on topographical maps, in records with meteorological data and surface water data, on soil maps, on geological maps and sections, and even on engineering maps. For many areas groundwater data is also directly available in the form of hydrogeological maps, water well records or groundwater reports. In order to check the information and to supply additional data, the desk

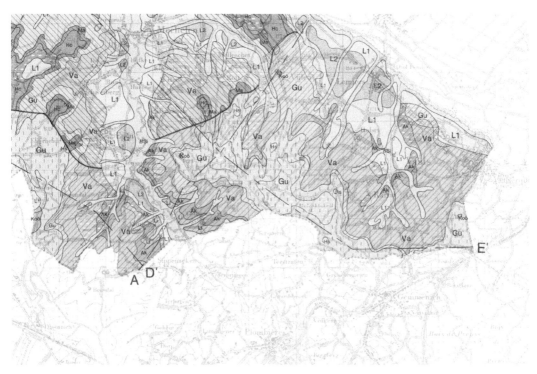

Figure 2.17. Part of a geological map of a sedimentary rock area. The Limburg Formation consisting of slate and quartzite is denoted by (Li), the sands and silts of the Aken and Vaals Formation by (Ak and Va), and the limestones of the Gulpen Formation by (Gu). Epen area, South Limburg, The Netherlands.

studies are usually combined with short visits to the field.

What kind of information on topographical and geological maps, in surface water records, etc. can be used for groundwater assessments? The following examples will show that these assessments can be manifold. For example, consider the topographical map. Areas with small villages or abundant vegetation may point to the occurrence of shallow groundwater tables or the presence of springs. Another item: the slope of the topography usually indicates the slope of the shallow groundwater tables. Directions perpendicular to the topographical contour lines give an indication of the direction of the flow of the shallow groundwater. Another example is the geological map. The rock types shown on the map may be used to set up a preliminary hydrogeological classification in terms of aquifers, aquitards, or aquifuges (Davis et al, 1967). The map therefore indicates characteristics of groundwater systems that may be present in an area (see section 2.1.2).

Figure 2.17 shows part of the Geological Map of South Limburg in The Netherlands (RGD, 1980). The rock types are predominantly consolidated and unconsolidated sediments, whereas small outcrops of metamorphic rock are also present. The main aquifers in the area include the unconsolidated fine sands of the Vaals and Aken Formations and the overlying semi-consolidated limestones of the Gulpen Formation. The metamorphic slates and quartzites of the Limburg Formation can be classified as an aquifuge acting as the impermeable base of the groundwater system.

Satellite imagery and aerial photography

Analysing satellite images and aerial photographs is partly done in the office and partly in the field. The fieldwork has to be undertaken to check the office interpretation of the images or photographs. Satellite images usually refer to the pictures taken by satellites. LANDSAT and SPOT images are available from respectively American and European satellites and lend themselves well for interpretation. In particular large geological complexes, major fault zones and regional folding can be well recognised on the images. A hydrogeological classification of the various rock types may be made.

A set of aerial photographs covers a much smaller area then a satellite image. The photographs are taken from a plane that traverses an area along parallel survey lines. The photographs partially overlap each other so that a three-dimensional view can be obtained when looking at them through a stereoscope. Aerial photographs lend themselves well for the topographical and geological interpretation of an area.

Figure 2.18 presents an example of an aerial photograph covering an area underlain by unconsolidated rock. The photo shows the physiography including a stream and a river forming part of the drainage system, a land use pattern consisting of agricultural land and forest area, and built up areas with the road network. Various landforms including dunes and ancient stream patterns can be interpreted on the basis of the three-dimensional view of the area. The eolian and alluvial character of the area points to an underlying unconsolidated groundwater system.

Hydrogeological mapping and well inventories

Hydrogeological mapping is a field activity that can also be combined with the fieldwork to be carried out for the interpretation of satellite images and aerial photographs. During the mapping in areas where the geology is exposed one inspects the prevailing rock types and tectonic features. In addition, an idea on the dimensions of the pores or fractures can be obtained and storage capacities and permeabilities of the rocks can be estimated. Mapping of the springs and seepage zones in an area confirms the occurrence of groundwater and gives an indication of the quality of groundwater (see section 6.3.2).

During well inventories existing wells in an area are visited. One can combine these surveys with the fieldwork to be carried out for hydrogeological mapping. Large diameter dug wells, single small diameter wells and well fields, and also infiltration galleries can be inspected. Data which are relevant for groundwater assessments may include rock types at the wells, depths at which water was struck, total

Figure 2.18. Aerial photo of an alluvial area covered with eolian sand (A) or river dunes (B). Minor (C) and main (D) streams and rivers are shown, as well as a well field (E) and a wastewater treatment plant (F). Doetinchem, The Netherlands (Topografische Dienst, 1997).

Table 2.11. Selected details of dug wells in the Appleby to Paynes Bay area, Barbados.

Name of well	Terrace	Lithology	Height surface (m.a.s.)	Total depth (m.b.s.)	Depth of water (m)	Chloride (mg/l)	In use?	Pump [1]
Branch, E	1	Limestone	2.1	2.3	0.20	552	Yes	W
Holder, L	1–2	Limestone	6.2	6.8	0.65	836	Yes	W
Hall, A	1–2	Limestone	5.9	6.4	0.55	424	Yes	H
Atwell, I	2	Limestone	9.9	10.3	0.50	752	Yes	W
Olive Lodge	2	Limestone	13.2	13.7	0.65	700	Yes	–

m.a.s. = metres above sea level. m.b.s. = metres below surface.
[1] W = Wind driven pump. H = Handpump.

depths of the wells, static groundwater levels in the wells, well yields, and water quality indicators. A common water quality indicator is the electrical conductivity (EC) which relates to the total dissolved solids (TDS) concentration in groundwater.

Table 2.11 shows an example of data collected during a well inventory in the Appleby to Paynes Bay area on the Island of Barbados (Senn, 1946). The table indicates basic characteristics of dug wells along the coast of this island in the Caribbean. The table shows that the wells were dug into the limestone aquifer to a depth just below sea level in order to prevent the intrusion of saline groundwater.

Geophysical surface investigations

Useful field investigations for assessing groundwater occurrences are geophysical surveys carried out at land surface. Several types of geophysical surveys can be distinguished: geo-electrical, electro-magnetic, magnetometric, seismic and gravimetric surveys. Most frequently applied in groundwater investigations are geo-electrical and electro-magnetic surveys (Nath et al, 2000).

Geo-electrical surveys make use of instrumentation whereby electric currents are injected into the ground and generated differences in potentials are measured. The electric currents are injected at steel or copper electrodes that are extended along a straight line. Figure 2.19

Figure 2.19. Execution of a geo-electrical measurement following the Schlumberger layout.

shows the execution of a geo-electrical measurement following the so-called 'Schlumberger layout'. The line of extension for the electrodes is indicated by the yellow tape. The shown instrument measures separately the injected current strengths and the generated potential differences. The small box shown in the picture contains batteries to provide the electrical current for injection.

During a geo-electrical measurement, the recorded current strengths and potential differences are recalculated into so-called 'apparent resistivities' that form the basis for interpretation. Nowadays, the interpretation of the apparent resistivities is carried out by personal computers. Computer codes interpret the apparent resistivities in terms of thicknesses and resistivities of individual subsurface layers. The correct interpretation of the resistivities leads to assessments on the rock type of the various layers and one may also make predictions on groundwater quality.

Figure 2.20 shows a computer interpretation of a geo-electrical measurement that was set up to investigate the groundwater system in the northeastern part of The Netherlands. The interpreted resistivities of the subsurface layers are in the order of 2 to 650 Ohm.m (say: ohmmeter).

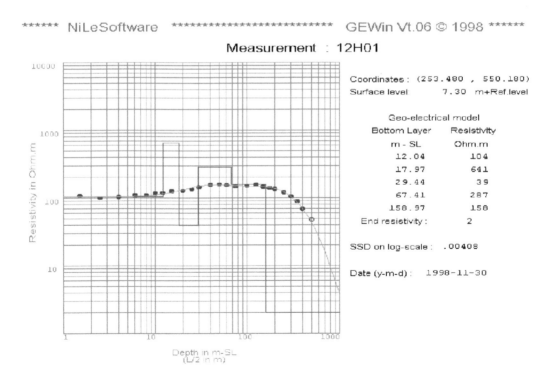

Figure 2.20. Field data and computerised interpretation of a geo-electrical measurement, Gasselte, The Netherlands (Bisht, 1999).

The low end resistivity of 2 Ohm.m points to brackish groundwater, whereas the resistivity in the order of 40 Ohm.m indicates a layer made up of loam or clay. The higher resistivities represent the series of saturated unconsolidated fluvial sands that constitute the main aquifers in the groundwater system.

Electro-magnetic surveys are much quicker done than geo-electrical surveys. The electro-magnetic method uses a magnetic field generated by a transmitter. A receiver registers a secondary magnetic field that is affected by subsurface rock properties including rock type, water content and water salinity. These properties can be estimated from the registration charts produced by the equipment. Since the computed apparent resistivities are usually not subjected to further computer interpretations, this method provides less information than a geo-electrical survey. However, the method is widely used for the identification of fractured zones in hard rock areas.

2.3.2 *Subsurface investigation methods*

Exploration drilling and logging

A number of sophisticated methods explore the occurrences and characteristics of groundwater systems below land surface. The most superior but also the most expensive field method that one can engage for

Figure 2.21. Direct (straight) circulation mud rotary drilling in the Sana'a Basin, Yemen.

subsurface groundwater investigations is exploration drilling. During this activity, a hole is drilled to target depth into the rock below land surface. Mainly depending on the type of rock in an investigation area, a suitable drilling method can be selected. Auger boring and cable tool drilling, direct or reverse circulation mud rotary drilling and air rotary percussion (down-the-hole hammer) drilling are widely used. Drilling techniques are not discussed in this textbook, but adequate information is available in a large number of texts on this subject (e.g. Driscoll, 1986).

Figure 2.21 shows truck-mounted equipment for the drilling of an exploration hole in volcanic aquiferous rock in Yemen. The drilling site was supposedly selected on a large fracture system in the rock. The selected drilling method is based on direct circulation mud rotary drilling with the assistance of compressed air. The high speed of the rotating drilling bit and the quick removal of rock fragments from the hole by the mud and the air guaranteed good drilling progress.

Rock sampling, geophysical logging and pumping tests (see section 3.2.4) are usually carried out during exploration drilling. Rock samples are taken when drilling is in progress. Geophysical logging in the borehole including so-called 'spontaneous potential (SP), long and short

Table 2.12. Table showing details on various drilling methods.

Drilling method	Well depth and well diameter	Suitable rock for drilling	Unsuitable rock for drilling	Typical penetration rates	Quality of rock sampling	Quality of geophysical logging	Quality of water sampling
Hand auger	Max. 10–15 m; 0.05–0.20 m.	Unconsolidated sediment	Hard rock	20–30 m/day	Good, hardly disturbed sample	Usually not carried out	Good, but mixed sample
Cable tool percussion	Max. 100–600 m; 0.1–0.9 m	Unconsolidated /consolidated sediment	Hard igneous/ metamorphic rock	3–30 m/day	Good, but disturbed sample	Usually not carried out	Good, but mixed sample
Mud rotary: Direct	Max 100–2000 m; 0.1–1.5 m	Most geological formations	Very hard rock/ cavernous rock/boulders	10–150 m/day	Rather poor	Logging required and usually satisfactory	Samples cannot be taken during drilling
Rotary: Reverse	Max 120–350 m; 0.4–1.8 m	Usually applied in unconsolidated formations	Hard rock/ cavernous rock/ boulders	50–100 m/day	Reasonable to good, but disturbed sample	Usually satisfactory	Drilling fluid will mix with formation water
Down the hole hammer	Depth depends on quantity of formation water encountered; 0.05–0.4 m	Hard to very hard rock	Unconsolidated sediments	20–75 m/day	Crushed, but reasonable sample	Satisfactory below water table	Good, but mixed sample

normal resistivity (LN and SN) and gamma logging' is carried out when drilling is interrupted or when drilling operations have been completed. During SP logging a potential difference, which is being created spontaneously in the drilled hole, is measured. In resistivity logging, the 'in-situ' resistivities of the various rock types are assessed. The gamma logging determines radiation in the subsurface that may reveal the presence of rocks containing isotopes, including clays and shales.

For an optimum interpretation, the results of rock sampling and geophysical logging activities are combined. This is usually done by plotting the various rock sample descriptions and geophysical logs against the same depth scale. This procedure enables one to make accurate lithological logs (columns with rock type symbols) and to evaluate water content and water quality information.

Table 2.12 has been prepared to relate the various drilling methods to the quality of rock sampling, geophysical logging and water sampling. In addition suitable rocks for drilling, maximum borehole depths and hole diameters are listed. The table indicates that drilling with methods whereby rock sampling is poor, i.e. direct circulation mud rotary drilling, needs to be followed by geophysical logging.

CHAPTER 3

Groundwater Movement

3.1 PRINCIPLES OF GROUNDWATER FLOW

3.1.1 *Groundwater flow in unconsolidated rock*

Darcy's Law

In the 19th century, the French water works engineer Henri Darcy published his now famous report on the movement or flow of groundwater in porous media (Darcy, 1856). He reported that he carried out experiments in the laboratory using sand columns. The columns were built in such a way that water could flow through these devices. Darcy found that the flow rate of water through the columns was related to the permeability of the type of sand used and the difference in height of the water levels at measuring points. The mathematical formulation of this relationship has led to the well-known Darcy's Law.

Figure 3.1 shows the principle of the experiment that Darcy carried out. The figure shows a circular column filled with sandy material, and water inflow and outflow tubes. Water is flowing through the device at a constant rate finding its path between the individual sand grains in the column. Water levels are measured in relation to an arbitrary reference level in two open tubes that are positioned at a fixed distance

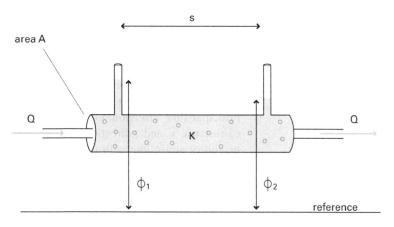

Figure 3.1. The Darcy laboratory set-up.

from each other. Darcy's Law for the set up shown can be expressed as follows:

$$Q = -AK \frac{\phi_2 - \phi_1}{s} \qquad (3.1)$$

where:
Q = flow rate through the sand column (m^3/day)
A = cross section of the sand column (m^2)
K = coefficient of permeability of the sand column (m/day)
ϕ_1, ϕ_2 = water levels in the tubes above reference (m)
s = distance between the tubes (m)

For the computation of groundwater movement or groundwater flow in natural groundwater systems, Darcy's Law is usually presented in a somewhat different form. The basic form of Darcy's Law is written in the differential form and the 'water levels in the tubes' are replaced by the so-called 'hydraulic heads' (see below). Instead of the flow rate through a cross section of a sand column, the 'flow rate through a unit area of rock material' is considered, also referred to as 'specific discharge'. Darcy's Law can then be formulated as follows:

$$q = -K \frac{d\phi}{ds} \qquad (3.2)$$

where:
q = flow rate through a unit area of $1m^2$, or specific discharge (m/day)
K = coefficient of permeability of rock material (m/day)
ϕ = hydraulic head (m)
s = distance measured in the direction of flow (m)

Different authors use different symbols and names for the parameters used in the equations for groundwater flow. In equation (3.2), the hydraulic head is introduced using the symbol ϕ. Other authors also use the symbol 'h' for this parameter. In some literature, the coefficient of permeability is also referred to as the hydraulic conductivity. The term $d\phi/ds$ is called the hydraulic gradient in differential form and in some textbooks is simply abbreviated by the character i.

The q in equation (3.2) is the flow rate through a unit area in the principal flow direction. One may also consider the flow rates in the cartesian x, y, and z directions. In case the principal directions of permeability are in line with the cartesian directions, then the following equations are valid:

$$q_x = -K_x \frac{\partial \phi}{\partial x} \qquad (3.3)$$

$$q_y = -K_y \frac{\partial \phi}{\partial y} \qquad\qquad (3.4)$$

$$q_z = -K_z \frac{\partial \phi}{\partial z} \qquad\qquad (3.5)$$

where:
q_x, q_y, q_z =   flow rates through a unit area of 1 m^2 in the x, y and z directions (m/day).
K_x, K_y, K_z = coefficients of permeability in the x, y and z directions (m/day).

Can Darcy's Law be used for any type of rock? The Darcy equation can be engaged to carry out computations of groundwater flow in groundwater systems consisting of unconsolidated rock, where pore spaces are present in between the rock grains. The flow of water takes place through these spaces. Whether Darcy's Law is valid in consolidated rock where pore spaces are usually absent and the flow is mainly through opened-up joints, faults, bedding plane contacts, solution holes or vesicles, is an issue that will be addressed in section 3.1.2.

The validity of Darcy's Law also relates to scale. In unconsolidated rock, Darcy's Law describes the flow rate through a unit area, i.e., a unit cross section through a block of rock. The flow rate can be considered as 'the flow rates through all the channels around the individual grains being lumped together on a 1 m^2 basis'. The equation does not describe the 'microscopic' behaviour of the flow rate through these channels. In other words, Darcy's Law falls short of describing the flow of groundwater on a microscopic scale.

Continuity equation

Darcy's Law is a powerful tool and can be used to compute groundwater flow rates in groundwater systems. Nevertheless, to solve a groundwater flow problem completely, the flow equation should be combined with the continuity equation. The continuity equation is the mathematical equivalent of the law of conservation of mass. To visualise the law of conservation of mass one can look at an elemental control volume as presented in Figure 3.2. The diagram shows the masses of groundwater that are entering or leaving one of the 'faces' of the control volume. The mass flow rate through a face is the product of the flow rate through a unit area, the surface area of the face and the groundwater density. One can express the mass flow rate through the left face as:

$$\rho q_x \, \Delta y \Delta z \qquad\qquad (3.6)$$

where:
ρ = groundwater density (kg/m^3)

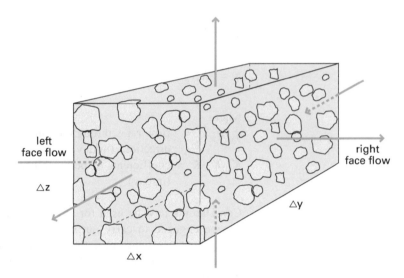

Figure 3.2. Elemental control
volume and mass flow.

$\Delta y, \Delta z =$ lengths of the faces of the control volume in the y and z
directions (m).

Subsequently, the mass flow rate through the right face can be
described by taking the first two terms of a Taylor series:

$$[\rho q_x + \frac{\partial(\rho q_x)}{\partial x} \Delta x]\Delta y \Delta z \qquad (3.7)$$

The change in mass flow rate in the x direction can then be found by
subtracting the equations (3.6) and (3.7):

$$-\frac{\partial(\rho q_x)}{\partial x}\Delta x \Delta y \Delta z \qquad (3.8)$$

The law of conservation of mass states the following. The sum of the
changes in mass flow rates in the x, y, and z directions is equal to the
mass rate of groundwater stored or released in the elemental control
volume. This law can now be expressed by the following relationship,
taking into account changes in mass flow rate in the form of equation
(3.8):

$$-[\frac{\partial(\rho q_x)}{\partial x} + \frac{\partial(\rho q_y)}{\partial y} + \frac{\partial(\rho q_z)}{\partial z}] \Delta x \Delta y \Delta z = \frac{\partial(\Delta M)}{\partial t} \qquad (3.9)$$

where:
$\Delta M =$ mass of groundwater in the control volume (kg).

The mass of groundwater, ΔM, can be expressed as the product of the volume of the elemental control volume, the water density and the porosity of the material. Equation (3.9) can then be written as follows:

$$- [\frac{\partial(\rho q_x)}{\partial x} + \frac{\partial(\rho q_y)}{\partial y} + \frac{\partial(\rho q_z)}{\partial z}] \Delta x \Delta y \Delta z = \frac{\partial(n\rho \Delta x \Delta y \Delta z)}{\partial t} \qquad (3.10)$$

where:
n = porosity of the rock material (dimensionless).

Equation (3.10) can be considered as the general continuity equation that complies with the law of conservation of mass. The equation can further be simplified for specific cases. First consider the case whereby one does not have any groundwater stored or released in the control volume itself. Equation (3.10) then reduces to:

$$\frac{\partial(\rho q_x)}{\partial x} + \frac{\partial(\rho q_y)}{\partial y} + \frac{\partial(\rho q_z)}{\partial z} = 0 \qquad (3.11)$$

In addition, assume that the water density, as a function of space, is constant. Equation (3.11) can then be worked out as the well-known continuity equation for steady, incompressible groundwater flow:

$$\frac{\partial q_x}{\partial x} + \frac{\partial q_y}{\partial y} + \frac{\partial q_z}{\partial z} = 0 \qquad (3.12)$$

Equation (3.12) may be combined with formulae expressing Darcy's Law: the formulae (3.3), (3.4) and (3.5). If one may assume that the rock material is homogeneous, then the following equation can be formulated:

$$K_x \frac{\partial^2 \phi}{\partial x^2} + K_y \frac{\partial^2 \phi}{\partial y^2} + K_z \frac{\partial^2 \phi}{\partial z^2} = 0 \qquad (3.13)$$

The rock may also be isotropic. Then equation (3.13) reduces to the well-known Laplace equation for steady flow:

$$\frac{\partial^2 \phi}{\partial x^2} + \frac{\partial^2 \phi}{\partial y^2} + \frac{\partial^2 \phi}{\partial z^2} = 0 \qquad (3.14)$$

Consider also the case whereby groundwater is stored or released in the control volume. The storage or release is mainly caused by variations in the time-related density of the groundwater itself, and the porosity of the rock material. By introducing the concept of specific storage, the mass rate of groundwater stored or released can be expressed in terms of a change in hydraulic head. Equation (3.10) can be taken as the starting point for the derivation of the relation, based on the concept of specific storage. However, in this textbook, the complete derivation will not be given. In case one assumes that the space-related density is constant, the derivation finally leads to the non-steady or transient continuity equation for groundwater flow:

$$-[\frac{\partial q_x}{\partial x} + \frac{\partial q_y}{\partial y} + \frac{\partial q_z}{\partial z}] = S_u \frac{\partial \phi}{\partial t} \qquad (3.15)$$

where:
S_u = specific storage; the volume of groundwater stored in, or released from a control volume of 1 m^3, for a 1 m increase or decrease of the hydraulic head (1/m).

Note that the continuity equation for non-steady flow transforms into the equation for steady flow when there are no changes in hydraulic head with time. The equation for non-steady flow may also be combined with the Darcy's Law equations. If one may assume that the rock is homogeneous, then the combination of equations (3.3), (3.4) and (3.5) with (3.15) yields:

$$K_x \frac{\partial^2 \phi}{\partial x^2} + K_y \frac{\partial^2 \phi}{\partial y^2} + K_z \frac{\partial^2 \phi}{\partial z^2} = S_u \frac{\partial \phi}{\partial t} \qquad (3.16)$$

The properties of the rock could point to an isotropic medium. Then one can rewrite equation (3.16) as follows:

$$\frac{\partial^2 \phi}{\partial x^2} + \frac{\partial^2 \phi}{\partial y^2} + \frac{\partial^2 \phi}{\partial z^2} = \frac{S_u}{K} \frac{\partial \phi}{\partial t} \qquad (3.17)$$

Equations (3.12), (3.14), (3.15) and (3.17) are considered the basic continuity equations for the description of groundwater flow. Naturally, the restrictions with regard to density and the coefficients of permeability that underlie these equations should not conflict with reality. The use of Darcy's Law and continuity equations for the computation of groundwater flow is extensively discussed in section 3.2 and other textbooks on this topic (e.g. Huisman, 1972; Fitts, 2002).

Permeability and groundwater flow

The permeability needs to be further discussed to get a better understanding of its meaning and role in the computations of groundwater flow. A mathematical expression for the coefficient of permeability was introduced as equation (2.2) in section 2.1.1. Experiments in the field and in the laboratory have shown that the intrinsic permeability in equation (2.2) can be evaluated. For unconsolidated rocks the following expression can be formulated:

$$k = Cd^2 \tag{3.18}$$

where:
C = shape factor (dimensionless)
d = average pore size between grains (m)

The shape factor C takes into account the effect of stratification of the grains, the packing of the grains, and the angularity of the grains. Equation (3.18) can be combined with equation (2.2) to yield the full expression for the coefficient of permeability in unconsolidated and other porous rocks:

$$K = Cd^2 \frac{\rho g}{\mu} \tag{3.19}$$

Another issue to be further reviewed in relation to permeability is the concept of 'isotropy and homogeneity'. This concept relates to the distribution of the coefficients of permeability in the different cartesian directions: the K_x, K_y and K_z. The concept can be described as follows:

– *Isotropy.* In a sequence of rocks as part of a groundwater system, the coefficients of permeability at particular locations may be the same in the *x*, *y* and *z* directions. In other words, the K_x, K_y and K_z are the same (see expressions (3.3), (3.4) and (3.5)). The rock is then said to be isotropic. On the other hand, the coefficients may differ in various directions: the K_x, K_y and K_z are not the same. Then, the rock is referred to as an anisotropic rock. For example, for a rock consisting of a very thin sub-horizontal layering of coarse sands and much finer material one can imagine that the coefficients of permeability taken in the horizontal direction are considerably larger than in the vertical direction. One can also say that the K_x and K_y are much larger than the K_z.

– *Homogeneity.* For a series of rocks, the coefficients of permeability may be the same 'from place to place'. One may also say that at different locations in the rock, the K_x, K_y and K_z are similar. The rock is then referred to as a homogeneous rock. Alternatively, the coefficients may differ from place to place: the K_x, K_y and K_z at one point are much different from the K_x, K_y, and K_z in another point. The rock is then said to be inhomogeneous.

An illustrative example of an anisotropic series of consolidated rocks is shown in Figure 3.3. The rocks are located in the Rada Basin in Yemen, which was introduced in section 1.2.5. The picture shows the

Figure 3.3. Picture showing the geological contact between sandstones (above) and meta-morphic basement rock (below) The sandstones which are aniso-tropic act as the local aquifer, Rada Basin, Yemen.

contact between the sedimentary sandstones and the metamorphic rock in the area. At this contact and further upward in the sandstones, at the contacts between the individual beds, the openings in the rock are mainly in a horizontal direction. In the saturated parts of this rock the horizontal coefficients of permeability, the K_x and K_y are larger than the K_z, and therefore the sandstone can be classified as anisotropic.

The concept of hydraulic head

The physical meaning of the hydraulic head in the Darcy's Law and continuity equations needs further explanation. The hydraulic head that can be defined in any location in a groundwater system is related to the so-called 'mechanical energy per unit mass of groundwater'. As illustrated in the left diagram of Figure 3.4, consider a point P in the groundwater system and the three types of 'mechanical energy' at this point. First, the 'gravitational energy' has to be considered. This is the energy required to lift a unit mass of groundwater from a reference level to point P. Secondly, the 'kinetic energy' should be taken into account. This is the energy needed to accelerate a unit mass of groundwater from 'standstill' to its actual velocity in the groundwater flow field at P. Third, the 'pressure energy' has to be taken into consideration. This is the energy required to raise the fluid pressure of a unit mass of groundwater from reference pressure to the fluid pressure at P.

From basic physics, one knows that the gravitational energy, for a unit mass, can be considered equal to:

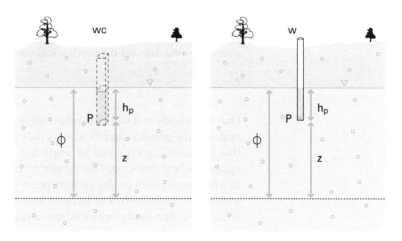

Figure 3.4. Cross section showing the elevation head (z) and the pressure head (h_p), for a water column (left) and for a well (right), at point P.

∘	sandy groundwater system	w	well with screen
▽	groundwater table	h_p	pressure head
.....	reference level	z	elevation head
wc	water column	φ	hydraulic head

$$E_e = gz \tag{3.20}$$

where:
E_e = gravitational energy (m^2/day^2)
g = acceleration of gravity (m/day^2)
z = distance from an arbitrary reference level to point P (m)

The kinetic energy can be expressed in terms of groundwater velocity. For a unit mass, the following expression holds:

$$E_k = \frac{v^2}{2} \tag{3.21}$$

where:
E_k = kinetic energy (m^2/day^2)
v = groundwater velocity (m/day)

Finally, the pressure energy can be written in terms of pressure and density. For a unit mass, the expression is:

$$E_p = \int_{p_0}^{p} \frac{dp}{\rho} \tag{3.22}$$

where:
E_p = pressure energy (m^2/day^2)
p = groundwater pressure at point P (kg/(m*day^2))
p_0 = reference pressure (kg/(m*day^2))

The mechanical energy per unit mass of groundwater is equivalent to the sum of the gravitational energy, the kinetic energy and the pressure energy. If one denotes the mechanical energy by E then:

$$E = gz + \frac{v^2}{2} + \int_{p_0}^{p} \frac{dp}{\rho} \tag{3.23}$$

One can modify and work out relationship (3.23). In most groundwater systems, the groundwater flow velocities are very small, which means that the kinetic energy can be neglected. Consider also groundwater systems where the water density can be considered constant and take as a reference pressure, the air pressure in the atmosphere to which one assigns the value zero. The integral in the expression for the pressure energy can then easily be solved and equation (3.23) transforms into:

$$E = gz + \frac{p}{\rho} \tag{3.24}$$

Equation (3.24) offers a pleasant expression for the mechanical energy, but is still not in a suitable form to define the hydraulic head. However, one may realise that the pressure at the point P can be written as:

$$p = \rho g h_p \tag{3.25}$$

where:
h_p = height of the water column at point P (m)

Combination of the equations, (3.24) and (3.25), yields the following expression for the mechanical energy:

$$E = gz + gh_p \tag{3.26}$$

The hydraulic head is now simply defined as the mechanical energy divided by the constant acceleration of gravity g:

$$\phi = z + h_p \tag{3.27}$$

The z and the h_p on the right side in equation (3.27) are also referred to as the 'elevation head' and the 'pressure head' at point P in the groundwater system. Equation (3.27) can also be expressed in terms of pressure. Combination of this equation with expression (3.25) yields:

$$\phi = z + \frac{p}{\rho g} \tag{3.28}$$

Measuring hydraulic heads

What can be concluded from the above? Perhaps the most important conclusion that can be drawn is that the defined hydraulic head has practical advantages. To make this clear, imagine the pressure head in an observation well with its screen at point P as shown in the right diagram of Figure 3.4. In the well, the pressure head is the height of the water column above the screen, which is equivalent to the distance from the screen to the groundwater level in the well. The elevation head is the distance from the screen to a reference level, usually mean sea level. The hydraulic head being the sum of the pressure head and the elevation head (equation 3.27) is then simply the distance from the groundwater level in the well to the reference level.

The hydraulic head can be determined from measurements at the observation well. Based on the considerations above, the following expression can be formulated:

$$\phi = Z - Z_d \tag{3.29}$$

where:
Z = distance from the top of the well to reference level (m)
Z_d = distance from the top of the well to the groundwater level (m)

The distances in equation (3.29) can easily be measured. The distance from the top of the well (top casing) to reference can be determined from topographical surveys including land surveying or altimeter surveys. The distance to the water level can be recorded with a water level gauge.

Hydraulic heads and pressure heads in a fresh-saline environment

Large differences in salinity may exist within a groundwater system. These differences also create variations in groundwater density across the system (see also section 2.1). Density variations may be so large that the hydraulic head, which was defined for a constant density in the previous section, has to be modified. Instead, for practical applications, pressure heads in saline groundwater are converted into equivalent pressure heads for fresh groundwater. The underlying assumption is that the pressures themselves do not change:

$$p_f = p_s \tag{3.30}$$

where:
p_s = groundwater pressure for saline groundwater (kg/(m*day^2))
p_f = groundwater pressure for fresh groundwater (kg/(m*day^2))

Inserting expressions similar to equation (3.25) into expression (3.30), the following expression is obtained:

$$\rho_f g h_{p(ef)} = \rho_s g h_{p(s)} \tag{3.31}$$

where:
ρ_f = density of fresh groundwater (kg/m^3)
ρ_s = density of saline groundwater (kg/m^3)
$h_{p(ef)}$ = equivalent pressure head for fresh groundwater (m)
$h_{p(s)}$ = pressure head for saline water (m)

Expression (3.31) can be re-arranged to find the equivalent pressure head for fresh groundwater. The elevation head which does not change in the conversion process, can also be added to find an expression for the equivalent hydraulic head:

$$\phi_{(ef)} = z + \frac{\rho_s}{\rho_f} h_{p(s)} \tag{3.32}$$

Expression (3.30) can also be considered to explain the findings of Badon Ghijben (1888) and Herzberg (1901). These researchers developed formulae for the computation of the depth to a sharp fresh-saline interface in a groundwater system. Figure 3.5 shows an example of an interface in a coastal area. At a point at the interface, the pressures in the adjoining fresh water and saline water environments are equivalent and expression (3.30) is valid. Expressions in the form of equation (3.25) can again be filled into (3.30), yielding the following relationship:

$$\rho_f g h_{p(if)} = \rho_s g h_{p(is)} \tag{3.33}$$

where:

$h_{p(if)}$ = pressure head for fresh groundwater at the interface (m)
$h_{p(is)}$ = pressure head for saline groundwater at the interface (m)

Figure 3.5 shows the pressure heads at the interface as measured in observation wells with the screens just above and below the interface. The pressure head for fresh groundwater can be split up in a part above the sea level and a part below this level. The part below the sea level is similar to the pressure head for the saline groundwater, assuming that no flow exists in the saline part of the groundwater system. In case of no flow there is hydrostatic equilibrium in the saline part and the pressure head at the interface is equal to the distance from the interface to sea level. Splitting up the pressure head for fresh groundwater in equation 3.33, and re-arranging terms yields:

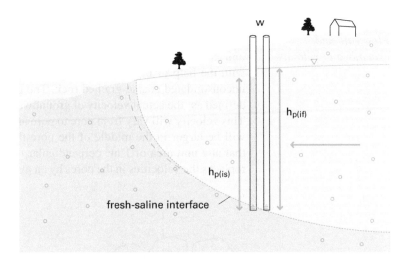

o	sandy groundwater system with fresh water
o	sandy groundwater system with saline water
▽	groundwater table
----	fresh-saline interface (Badon Ghijben)
--	fresh-saline interface (actual, near sea bottom)
←	groundwater flow
$h_{p(if)}$	pressure head for fresh groundwater
$h_{p(is)}$	pressure head for saline groundwater
w	observation wells

Figure 3.5. Schematic cross section of fresh groundwater overlying saline water, showing the sharp fresh-saline interface.

$$h_{p(is)} = \frac{\rho_f}{\rho_s - \rho_f} \Delta h \qquad (3.34)$$

where:

Δh = part of thé pressure head for fresh groundwater above the sea level (m)

Using equation (3.34), Badon Ghijben and Herzberg assessed the depth to the interface below sea level, which is the same as the computed pressure head for the saline groundwater. Equation (3.34) shows that the pressure head for the saline groundwater at the interface relates to the groundwater densities and the pressure head of fresh groundwater above the sea level. The densities for fresh and saline groundwater are generally known or can be measured. The pressure head above sea level can be estimated from groundwater level observations in shallow wells, just tapping the groundwater table. By inserting the values of these parameters into equation (3.34), the depth to the interface, from sea level, can be computed.

Flow rate and groundwater velocity

What is the relation between the groundwater flow rate through a unit area and groundwater velocity? This relation will be illustrated with the help of Figure 3.6 showing the pore space and grains in an unconsolidated coarse-grained rock. The groundwater velocity can be defined as 'the actual velocity of groundwater in the pores of the rock'. This velocity will vary from pore to pore and within a pore the velocity will be larger in the middle of the pore than near the grain. Imagine that at a unit area of 1 m^2 perpendicular to the direction of flow, one replaces the velocities in the pores by an average groundwater velocity.

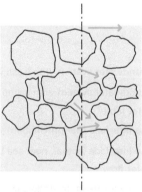

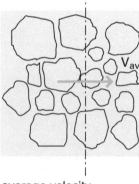

Figure 3.6. Groundwater flow velocities in unconsolidated rock.

actual velocities in pores

$\cdots$— projection of unit area

average velocity

The flow rate through 1 m^2 of rock can then be set equal to 'the average groundwater velocity multiplied by the open area of the pores'. If one assumes that the open area of the pores at the unit area can be approximated by the porosity, then the following relationship holds:

$$q = n v_{av} \qquad (3.35)$$

where:
q = flow rate through a unit area of 1 m^2 (m/day)
n = porosity (dimensionless)
v_{av} = average groundwater flow velocity (m/day).

The porosity in equation (3.35) may be replaced by the so-called effective porosity. In particular in fine-grained unconsolidated rocks including fine sands, silts and clays, groundwater in isolated pores and groundwater 'stuck' to the grains will not contribute to the flow of groundwater. The active pore space where groundwater flow takes place, the effective pore space, is smaller than the total pore space. Within this context, the effective porosity could be defined as 'the ratio of the volume of effective pore space and the total volume of rock'.

Equation (3.35) can be re-arranged and combined with the Darcy's Law equation (3.2) to find a further expression for the average groundwater velocity:

$$v_{av} = \frac{q}{n} = - \frac{K}{n} \frac{d\phi}{ds} \qquad (3.36)$$

Equation (3.36) relates the average groundwater flow velocity to the coefficient of permeability, the porosity and the hydraulic gradient in differential form. Since velocity and travel time are related, the equation can also be used as a basis for the computation of the time needed for groundwater to travel through (parts of) unconsolidated groundwater systems.

3.1.2 *Groundwater flow in consolidated rock*

Flow in fractures and solution openings

Can one apply the groundwater flow equations for unconsolidated rocks also to consolidated rocks? In section 3.1.1, the Darcy's Law and continuity equations have been discussed. Without reservation, these equations can also be applied to consolidated rocks where the flow of groundwater takes place in the pore space between the grains. Primary porosity is dominant in these rocks. Non-cemented sandstone is an example of a rock where primary porosity plays an important role. In most of the consolidated rocks, however, the open spaces are present at opened-up fractures, at bedding planes contacts, and at solution holes. Secondary porosity is prevailing. The question remains whether the Darcy's Law and continuity equations can still be used for these rocks.

Before answering this question one should understand the character of the flow of groundwater taking place in consolidated rocks with a predominantly secondary porosity. First, one will have to realise that in these rocks the flow of groundwater has a rather uneven distribution. In the more permeable parts with the openings, the preferred passageways for flow are located. In other parts of the rock with much less open space, the flow can be rather insignificant or there may even be no groundwater flow at all. Secondly, one expects that the magnitude of the flow depends on the widths of- and the inter-connection between the open spaces. The flow of groundwater will be larger in the wider- and better-connected spaces.

In view of the above, two approaches may be followed to formulate groundwater flow equations in consolidated rocks with secondary porosity. A first approach is to set up separate sets of flow equations for individual networks of opened-up fractures or solution holes, and the rock mass. This approach can be referred to as a 'non-continuum' approach. In the second approach the consolidated rock is replaced with a representative homogeneous rock to which averaged values are assigned for the porosity, the specific storage, and the coefficient of permeability. Groundwater flow equations for the 'replacement' rock are then used which are similar to the flow equations derived in section 3.1.1. This second approach is called the 'continuum' approach.

Where will one use the non-continuum or continuum approach? In this textbook the emphasis will be on the continuum approach. This approach can be used for groundwater systems, which have an even distribution of opened-up fractures and other types of open spaces in the rock mass. Also, the openings should be well connected over larger areas. In addition Darcy's Law should be valid (see below). Large-scale groundwater flow assessments may also favour the selection of the continuum approach. For example, for groundwater flow computations carried out for a large regional groundwater system, the continuum approach may be considered.

Validity of Darcy's Law   The most intriguing issue concerns the applicability of the Darcy's Law equation for consolidated rocks, where secondary porosity is a dominant feature. The Darcy's Law as presented in equation (3.2) assumes that the flow is 'laminar' and shows that the relationship between the hydraulic gradient and the flow rate per unit area is linear. Laminar groundwater flow means that 'the water moves in a sheet like fashion'. There may also be non-linear laminar flow, and flow whereby groundwater 'moves in a curl like fashion'. This latter type of flow is referred to as 'turbulent' flow. For non-linear laminar and turbulent flow, there are no linear relationships between the hydraulic gradient and the flow rate per unit area, and the Darcy's Law equation takes on a different form:

$$q = -K \left[\frac{d\phi}{ds} \right]^m \qquad (3.37)$$

where:
m = coefficient, less than 1 for non-linear laminar and turbulent flow

The so-called 'Reynolds number' can be used as a criterion to assess whether groundwater is in a linear and laminar, non-linear laminar or rather in a turbulent state. The Reynolds number (R_e) has been defined as follows:

$$R_e = \frac{\rho d q}{\mu} \qquad (3.38)$$

where:
ρ = density of groundwater (kg/m^3)
d = characteristic width of an opened-up joint, fault or other opening (m).
q = flow rate per unit area of 1 m^2 (m/day)
μ = dynamic viscosity (kg/(m*day))

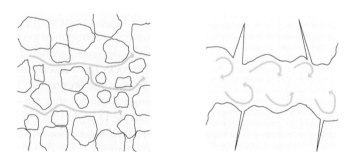

Reynolds number

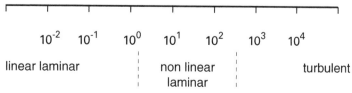

Figure 3.7. Diagrams showing the behaviour of groundwater flow and the related ranges for the Reynolds number.

Groundwater flow is linear and laminar in those parts of a rock where the Reynolds number is below some value in the range of 1 to 10. The flow is non-linear laminar when this number takes on values in the range of, say, 10 to 100. Turbulent flow is predominant when the Reynolds number is above 100. Figure 3.7 shows these ranges in an illustrative diagram. In case one is able to indicate values for the Reynolds number, using rough estimates on flow rates and widths of opened-up joints and faults, then the character of the flow of groundwater can be predicted, and the validity of Darcy's Law can be assessed.

Flow characteristics for various consolidated rocks

One can visualise the flow characteristics in different types of consolidated rock with secondary porosity. This means that an effort can be made to predict whether a continuum approach can be followed, and the Darcy's Law equation can be used in its original form. A tentative outline on the flow characteristics of the consolidated rock types, as introduced in section 2.2 is as follows:

– *Metamorphic and intrusive rock.* Generally, these rocks including gneisses, schists, quartzites, granites, and diorites have a very low permeability. However, several systems of opened-up fractures may have developed in the upper parts of these rocks or in the weathered zones. Local aquifers may have formed for which groundwater flow computations may be carried out. In case the distribution of the connected fractures is rather dense and their widths are not excessive then a continuum approach and the Darcy's Law equation could be applied for the local aquifers.

– *Volcanic rock.* These rocks, including basalts, rhyolites and tuffs, often show a large variation in porosity and permeability. Aquifer zones consisting of fractured rock, layers with vesicles and sedimentary intervals, are intercalated with aquifuges made up of dense impermeable rock. The uneven distribution of the porosity in volcanic rock and the possible presence of large separated networks of openings make it doubtful whether, for the rock as a whole, the continuum approach and the Darcy's Law equation can be applied.

– *Consolidated sediments.* For these rock types, one will have to make a distinction between the fine- and coarse-grained sedimentary rocks, and the carbonate rocks. The fine-grained consolidated sediments including shales, siltstones, and mudstones are generally considered impermeable rocks, but they may locally be permeable at fractures. If the distribution of connected fractures is dense, then a continuum approach and the Darcy's Law equation could be used to describe, just locally, the movement of groundwater.

Coarse-grained consolidated sediments include sandstones and conglomerates. In case these rocks are well cemented and alternated with fine-grained sediments, then their permeabilities are generally low.

Other sandstones may be quite permeable as a result of open pore space (primary porosity) and the occurrence of open space associated with joints, faults, and bedding plane contacts. In these permeable sediments, acting as aquifers, the continuum approach could be used and the Darcy's Law equation can be applied.

Consolidated carbonate rocks comprise limestones and dolomites. Carbonate rocks and volcanic rocks may show similarities in a hydraulic sense. Perhaps, the unbalanced distribution of porosity and permeability in the 'carbonates' is even more outspoken than in the 'volcanics'. Large porosities in the carbonate rocks are primarily a result of the formation of solution channels, although open space may also have developed at fractures or at contacts between bedding planes. In particular in the larger solution channels the assumption of linear and laminar flow will not be valid. In many areas with carbonate rocks, it will be highly questionable whether the continuum approach and the Darcy's Law equation can be used.

3.2 FLOW IN GROUNDWATER SYSTEMS

3.2.1 *Basic concepts*

Flow schematisation

How does one apply the basic groundwater flow equations? In section 3.1, the Darcy's Law and continuity equations have been introduced as a basis for the computation of groundwater flow. The Darcy's Law equation can be applied directly, but one may also follow an analytical approach whereby the combined Darcy's Law and continuity equations are solved, taking into account realistic boundary conditions. Yet another approach is the numerical treatment of the basic equations, which forms the platform for the building of groundwater models. However, whether opting for a direct, an analytical or for a numerical solution, one will have to realise the 'dimensionality' of the flow in the groundwater system that one wishes to consider.

Groundwater flow in groundwater systems is nearly always three-dimensional. For example, one will understand that the flow of groundwater originating from precipitation which recharges a groundwater system, then flows through the system and discharges at springs, streams and rivers, is essentially three-dimensional. However, groundwater systems are usually simplified and sub-divided so that 'horizontal and vertical' flow computations can be carried out.

Groundwater systems consist of intricate systems of aquifers, aquitards and aquicludes (see section 2.2). In large parts of these systems, the flow in the aquifers can be assumed as horizontal and the flow in the aquitards as vertical. The assumption of horizontal flow in the aquifers

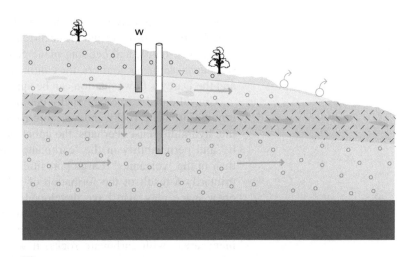

Figure 3.8. Cross section showing groundwater flow in an extensive groundwater system.

⬟ aquitard		▽ groundwater table	
unconfined aquifer		→ groundwater flow	
○ semi-confined aquifer		w well with screen at bottom	
⬛ aquifuge		ᕤ springs	

is also referred to as the 'Dupuit assumption' (1863). A schematisation into horizontal flow in the aquifers and vertical flow in the aquitards is especially allowed when the lateral extents of these units are large in comparison with their thicknesses. In Figure 3.8, this schematisation is illustrated. In the cross section presenting an extensive groundwater system, horizontal flow is indicated in the aquifers and vertical flow is shown in the aquitard.

Transmissivity, vertical resistance and storage

For groundwater systems where the schematisation of horizontal flow in the aquifers and vertical flow in the aquitards is allowed, the following definitions can be introduced. The transmissivity of an aquifer in the groundwater system is defined as 'the product of the horizontal coefficient of permeability and the saturated thickness of the aquifer'. In case one assumes that the aquifer is isotropic ($K_x = K_y = K$; see section 3.1.1) then one can write for the transmissivity:

$$T = K H \tag{3.39}$$

where:
T = aquifer transmissivity (m²/day)
H = saturated aquifer thickness (m)

The vertical resistance of an aquitard is defined as 'the ratio of the thickness of the aquitard and its permeability in the vertical direction (K_z)':

$$c = \frac{D}{K_z}$$

(3.40)

where:
c = vertical resistance (days)
D = thickness of the aquitard (m)

The specific storage has been defined in section 3.1.1 as 'the volume of water stored in, or released from a control volume of 1 m³, for a 1 m increase or decrease in hydraulic head'. Figure 3.9 illustrates this control volume in a cross section through unconsolidated rock. For confined and semi-confined aquifers one can define the storativity as 'the volume of water stored in, or released from a control volume, consisting of an aquifer column with a cross sectional area of 1 m², for a 1 m increase or decrease in hydraulic head'. Figure 3.9 also shows an aquifer column in a confined or semi-confined aquifer. In these aquifers, water is stored or released, mainly due to elastic changes in porosity and groundwater density. These changes are a result of the changes in pressures that are exerted on the aquifer. Since the underlying assumptions for the storage and release mechanisms are similar, the storativity can be related to the specific storage as follows:

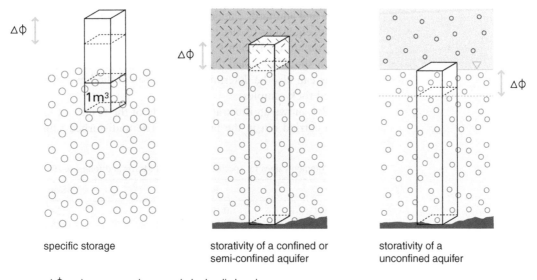

specific storage   storativity of a confined or   storativity of a
semi-confined aquifer   unconfined aquifer

Δφ   increase or decrease in hydraulic head

▽   groundwater table

Figure 3.9. Cross sections showing the concept of storage in unconsolidated aquifers.

$$S = S_u H \tag{3.41}$$

where:
S = storativity (dimensionless)
H = aquifer thickness (m)

For unconfined aquifers, the storativity can be defined as 'the volume of water stored in, or released from a control volume, made up of an unconfined aquifer column with a cross sectional area of 1 m², as a result of a 1 m increase or decrease in hydraulic head'. Figure 3.9 also presents an aquifer column in an unconfined aquifer. The storativity in an unconfined aquifer is nearly completely made up of the so-called specific yield, S_y, which relates to the water that can be drained from pore space. The 'working mechanism of storage' in an unconfined aquifer is different from the mechanism in confined and semi-confined aquifers. In an unconfined aquifer, water is not primarily stored or released as a result of changes in porosity and density. Groundwater is mainly stored or released because the pore space is 'filled or drained' at the groundwater table.

Laboratory work   How does one determine transmissivities, vertical resistances, storativities and specific yields? Values for these parameters can best be determined from pumping tests to be carried out in the field. These tests are discussed in more detail in section 3.2.4. In case the thicknesses of aquifers and aquitards can be established from geophysical surveys and drilling, equations (3.39), (3.40) and (3.41) can be re-arranged and used to compute coefficients of permeability and specific storages. However, it should be noted that coefficients of permeability and specific yields can also be determined from tests in the laboratory using undisturbed rock samples.

Table 3.1 shows an illustrative example of the results of tests that have been carried out in the laboratory. The undisturbed samples of

Table 3.1. Common values for specific yields (after Morris and Johnson, 1967).

Type of rock	Range	Mean
Medium gravel	0.17 – 0.44	0.24
Fine gravel	0.13 – 0.40	0.28
Medium sand	0.16 – 0.46	0.32
Fine sand	0.01 – 0.46	0.33
Silt	0.01 – 0.39	0.20
Clay	0.01 – 0.18	0.06
Tuff	0.02 – 0.47	0.21
Sandstone	0.02 – 0.30	0.21
Sandstone	0.12 – 0.30	0.27
Siltstone	0.01 – 0.28	0.12

a variety of sedimentary rocks were impregnated with water and the loss of water from the samples by gravity was measured. This enabled the researchers to compute specific yield values and the table shows ranges and mean values. The table indicates that mean specific yields generally vary between 0.20 and 0.35, with the exception of clays and siltstones which have substantially lower values for this parameter.

Equipotential surfaces and flow directions

What definitions relating to hydraulic heads in groundwater systems can be formulated? First, definitions and concepts concerning the distribution of hydraulic heads in groundwater systems, in general, will be discussed. For a system, equipotential surfaces can be defined as 'surfaces where the hydraulic heads all have the same value'. For isotropic conditions, it can be shown that the flow of groundwater is then perpendicular to these equipotential surfaces, flowing from places with higher hydraulic heads to places with lower heads. Thus, one may conclude that the flow directions in an isotropic groundwater system can be determined when these equipotential surfaces have been established. However, this conclusion can also be turned the other way around. The orientations of the surfaces can be established when the groundwater flow directions are known.

The outlined definitions may be used for groundwater systems where the horizontal and vertical schematisation of groundwater flow is allowed. In the aquifers, where the groundwater flow is assumed horizontal, the orientation of the equipotential surfaces of constant hydraulic heads can be defined for isotropic conditions. Being perpendicular to groundwater flow, these surfaces are vertical. In the aquitards where the flow is considered vertical, the equipotential surfaces are horizontal. The particular features of the equipotential surfaces in groundwater systems will be one of the cornerstones for the groundwater flow computations that will be elaborated in the next sections.

3.2.2 *Regional groundwater flow*

Regional flow and groundwater head contour maps

Regional groundwater flow can be considered as 'the flow through large parts of a groundwater system'. Within the context of this textbook, the main focus is on the analyses of regional flow. The emphasis is less on the local flow of groundwater including the flow to canals, to wells, to building pits, to mines, etc. (see section 3.2.3). Regional flow can be analysed using the basic groundwater flow equations. The methods for the computation of regional groundwater flow to be discussed in the next sections generally assume horizontal flow in the aquifers and vertical flow in the aquitards. The methods are the following:
– Methods based on *simple calculations*.
– Methods focusing on the compilation of *flownets*.
– Methods based on the application of *groundwater models*.

Whatever method is followed for the flow computations, the compilation of groundwater head contour maps is a crucial activity. Groundwater head contour maps are maps that show projections of the equipotential surfaces of the hydraulic heads. Usually, the maps present horizontal projections of the equipotential surfaces of the hydraulic heads which are representative for the aquifers in a groundwater system. Generally, these maps are also referred to as 'groundwater level contour maps'. More specifically, the term 'phreatic groundwater level contour maps' is in use for unconfined aquifers, and the term 'piezometric groundwater level contour maps' is employed for semi-confined and confined aquifers.

The maps can be prepared in a simple way. Based on measured distances to groundwater levels in selected observation wells, the hydraulic heads can be computed in metres above reference which is usually taken as mean sea level (see equation 3.29). Then, for relevant aquifers, the hydraulic heads can be plotted on suitable base maps. By contouring the heads, the groundwater head contour maps are obtained. On the maps, groundwater flow directions may also be indicated, perpendicular to the contour lines, in case it can be assumed that the aquifers are isotropic.

Figure 3.10 shows an illustrative example of groundwater head contour maps for a groundwater system near the city of Zhengzhou in China (MGMR & TNO, 1989). The groundwater system consists of a shallow and a deep aquifer separated by an aquitard. The flow in the aquifers is nearly horizontal, whereas the flow in the aquitard is roughly vertical. For July 1984, the maps show the groundwater head contour lines with reference to mean sea level. One map shows the contours of the shallow aquifer and the other map presents the contours of the deep aquifer. It is indicated that the hydraulic heads in the shallow aquifer are slightly higher than the heads for the deep aquifer. The maps also show the groundwater flow directions. For the shallow aquifer, a general flow direction towards the Yellow River is shown. For the deep aquifer, it is indicated that the direction of groundwater flow is towards the area with depressed hydraulic heads near the town of Zhengzhou. The reason for the flow towards the area with depressed hydraulic heads in the deep aquifer is the excessive pumping for the town.

Simple computations for the flow in aquifers

How can one estimate the flow of groundwater in a groundwater system using simple computations? First consider the flow in an aquifer and assume that this flow is horizontal. To find an expression for the computation of the flow rate through the aquifer, Darcy's Law can be considered (equation 3.2). Instead of taking the flow rate through a unit area of 1 m^2, the flow rate through the cross sectional area of the aquifer will be evaluated. In addition, the differential form of the

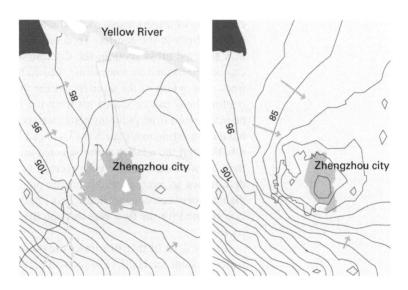

Figure 3.10. Groundwater head contour maps of the shallow phreatic aquifer (left map) and the deep semi-confined aquifer (right map), Zhengzhou city, China.

<u>105</u>   groundwater head contour line

⟶   flow direction

hydraulic gradient is not considered. For the gradient, a discrete difference in hydraulic head over a discrete distance will be taken. The flow rate through the aquifer at a selected cross section can then be computed as follows:

$$Q = -KHW\frac{\Delta\phi}{\Delta s} \tag{3.42}$$

where:
Q = groundwater flow rate through the aquifer (m³/day)
K = horizontal coefficient of permeability (m/day)
H = saturated thickness of the aquifer (m)
W = width of the aquifer (m)
$\Delta\phi$ = discrete difference in hydraulic head (m)
Δs = discrete distance (m)

Equation (3.42) can also be combined with expression (3.39). The resulting equation is as follows:

$$Q = -TW\frac{\Delta\phi}{\Delta s} \tag{3.43}$$

Figure 3.11 explains the use of equation (3.43). The figure presents a map showing a sandy permeable aquifer bordered by an aquifuge consisting of impermeable rock. The map shows the groundwater head contour lines in the aquifer. To compute the flow through the aquifer at a selected cross section, the discrete difference in hydraulic head can be determined by subtracting the heads for the two nearest contour lines. The width of the aquifer and the discrete distance between the contour lines can be scaled off from the map. The transmissivity can be interpreted from pumping tests (see section 3.2.4). When the values have been estimated, equation (3.43) can be used to compute the flow rate through the aquifer at the selected cross section. It is emphasised that one works in practice with average values for the parameters. One determines an average width for the aquifer and an average discrete distance between contour lines. This implies that also an average value is computed for the flow rate through the aquifer at the selected cross section.

Figure 3.12 further illustrates the outlined procedure for a groundwater system in Haiti (Euroconsult, 1989). The figure presents a groundwater head contour map drawn up for a river valley area near Gonaives. The local groundwater system consists of a single aquifer made up of sandy material. The system is bounded by an aquifuge of impermeable marly limestones. The flow rate through the aquifer at a cross section in the middle part of the river valley can be computed. For the computation, the two adjoining contour lines of 12 and 16 m above mean sea level can be considered. The discrete difference in hydraulic head is then 4 m. The average width of about 9000 m for the aquifer at the two contour lines can be scaled off from the map. The average discrete distance between the contour lines is around 1200 m. Assume that a value for the

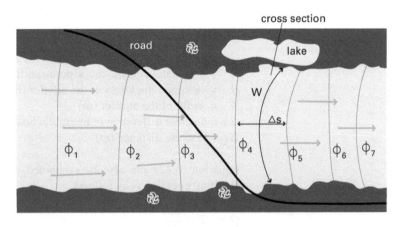

Figure 3.11. Map showing an aquifer and hydraulic head contour lines for the computation of the flow in the aquifer.

aquifer	W	width of aquifer
aquifuge	ϕ_1	groundwater head contour line
→ groundwater flow direction	Δs	distance between contour line

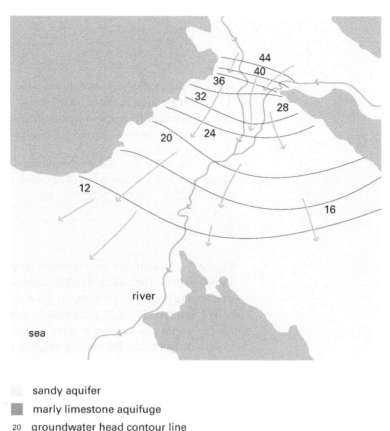

Figure 3.12. Groundwater head contour map for a confined aquifer near the town of Gonaives in Haiti, Caribbean.

 sandy aquifer

 marly limestone aquifuge

20 groundwater head contour line

→ groundwater flow

transmissivity of 1000 m^2/day can be evaluated from pumping tests. With all the parameters known, equation (3.43) can be used to estimate, at the selected cross section, a flow rate of 30,000 m^3/day through the aquifer.

Simple computations for the flow through aquitards

The estimation of the flow of groundwater through an aquitard can also be considered. Assume that this flow is vertical. To formulate an expression for the computation of the flow rate through the aquitard, one can take Darcy's Law in the vertical (z) direction as a starting point (equation 3.5). Instead of considering the flow rate through 1 m^2 of aquitard, the flow rate through the whole aquitard with a specified surface area will be considered. Also, the differential form for the hydraulic gradient in the vertical direction will not be taken, but a discrete difference in hydraulic head and a discrete distance across the aquitard will be considered. Equation (3.5) can then be written as:

$$Q_z = -K_z \, A \, \frac{\Delta \phi}{\Delta z} \qquad (3.44)$$

where:
Q_z = groundwater flow rate through the aquitard (m³/day)
A = surface area of the aquitard (m²)
$\Delta \phi$ = discrete difference in hydraulic head across the aquitard (m)
Δz = discrete distance across the aquitard (m)

The discrete distance across the aquitard is the same as the thickness of the aquitard. Combination of equation (3.44) with expression (3.40) then yields:

$$Q_z = -A \, \frac{\Delta \phi}{c} \qquad (3.45)$$

Figure 3.13 illustrates the computation procedure. The cross section shown indicates an aquitard sandwiched between two aquifers. In order to compute the flow rate through the aquitard, the discrete differences in hydraulic heads can be estimated by considering the hydraulic heads for the upper and the lower aquifer. Differences in heads estimated at sets of wells can be used to prepare a map with head differences for both aquifers. The surface area of the aquitard can also be scaled off from a map. In addition, resistances of the aquitard can be evaluated from pumping tests. When the values have been assessed, equation (3.45) can be used to compute the vertical flow rate through the aquitard.

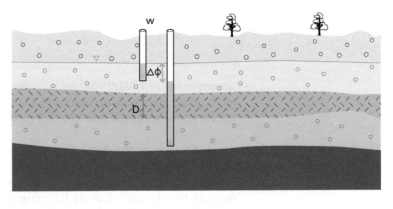

Figure 3.13. Cross section indicating the main parameters for the computation of the vertical flow through an aquitard.

○	unconfined aquifer	▽	groundwater table
○	semi-confined aquifer	Δϕ	difference in hydraulic head
■	aquifuge	D	thickness of aquitard
⟋	aquitard	w	well with screen at bottom

It is stressed that in the process one usually determines, for the aquitard area covered, average values for the parameters. Thus, an average value for the head differences for both aquifers, and an average resis-value for the aquitard are determined. This leads to the compution of an average flow rate through the aquitard. One should realise that locally this flow rate may deviate substantially from the computed average value for the aquitard.

Simple computations for groundwater storage

What can be said about the volumetric rates of groundwater stored or released in a groundwater system? Volumetric rates of water stored or released in an aquifer are usually larger than the rates for an aquitard. In order to compute these rates for an aquifer, the right side of the continuity equation for non-steady flow (equation 3.15) can be taken as a guide. Instead of assuming a rate for a small cross sectional area, the rate for the aquifer with a specified surface area will be considered. In addition, a discrete difference in hydraulic head over a discrete length of time, a time period, will be taken. Volumetric rates for confined or semi-confined aquifers will first be evaluated. For the computation of rates of water stored or released, over the whole aquifer thickness, the storativity should be taken into account (see section 3.1.2). The rates of water stored or released can be computed as follows:

$$S_{gws} = BS \frac{\Delta \phi}{\Delta t} \tag{3.46}$$

where:
S_{gws} = volumetric rate of groundwater stored or released (m^3/day)
B = surface area of the aquifer (m^2)
S = storativity (dimensionless)
$\Delta \phi$ = discrete difference in hydraulic head (head as piezometric groundwater level) for Δt (m)
Δt = discrete length of time (days)

Second, the rates for an unconfined aquifer can be considered. For this type of aquifer, the specific yield, S_y, has been introduced. The expression for the computation of volumetric rates of stored or released groundwater is as follows:

$$S_{gws} = BS_y \frac{\Delta \phi}{\Delta t} \tag{3.47}$$

where:
$\Delta \phi$ = discrete difference in hydraulic head (head as groundwater table elevation) for Δt (m)

How can one make use of the above formulations for the computa-tion of volumetric rates of water stored or released in aquifers? The

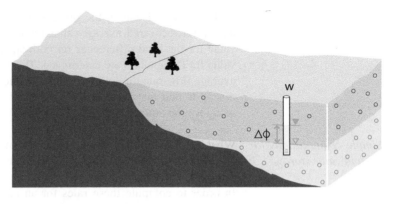

○	unconfined aquifer
■	aquifuge
▼	hydraulic head at start time period
▽	hydraulic head at end period
w	well with screen at bottom
Δφ	difference in hydraulic head

Figure 3.14. Block diagram indicating an unconfined aquifer and part of the parameters for the computation of released volumes.

procedures for semi-confined, confined, or unconfined aquifers are similar. Figure 3.14 illustrates the computation for an unconfined sandy aquifer shown in a block diagram. The single aquifer is underlain by an aquifuge of impermeable rock. The computation does not necessarily have to be performed for the whole aquifer, but the part of the aquifer, shown in the diagram, could also be considered. To perform the computations, discrete differences in hydraulic heads can be determined by taking into account the heads at the end and the heads at the start of a time period. Differences in head measured at observation wells during the time period can be used to assemble a map with head differences. The surface area can be scaled off from the maps. The value for the specific yield should preferably be assessed from pumping tests. When all the values have been determined, equation (3.47) can be used to compute the volumetric rate of water released in the unconfined aquifer.

Flownet analysis

Groundwater flow computations can be carried out effectively with flownets. Flownets are prepared, by drawing on groundwater head contour maps, a set of groundwater flow lines perpendicular to the contour lines. This can be done when conditions for isotropy are satisfied. One will find that the end result is a map showing a pattern of squares and rectangles. Such a grid is referred to as a flownet. Usually, flownets are prepared for aquifers where the assumption of horizontal flow is allowed.

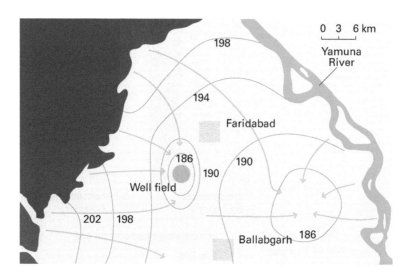

Figure 3.15. Map showing the alluvial groundwater system and the sketched interpreted flownet in the Faridabad area, south of New Delhi, India.

    sandy deposits

■   quartzite

190 groundwater head contour line

→ flow line

Figure 3.15 shows a flownet covering an area of alluvial sandy deposits in the valley of the Yamuna river in Central India (CGWB, 1989). The sandy deposits surrounded by impermeable quartzite form the main aquifer in the extensive groundwater system. The map shows the contour lines for the hydraulic heads and the perpendicular flowlines. They form a rather complex flownet indicating the flow of groundwater to a well field near Faridabad and to another area near Ballabgarh. The flow towards this latter area might have been induced by groundwater abstractions for irrigation schemes.

Flownets have a characteristic shape at the boundaries of a groundwater system. Figure 3.16 shows the flownets at three common types of boundaries: impermeable boundaries, constant head boundaries and groundwater table boundaries. The following comments can be made:

– *Impermeable boundary*: This boundary includes the contact between a groundwater system and impermeable rock. Since there can be no flow across the boundary, flow lines can only run parallel to the boundary. The contour lines for the hydraulic heads are then perpendicular to this boundary.

– *Constant head boundary*: This boundary could be the contact zone between a groundwater system and a surface water system including a sea, lake, stream or river. If one may assume that the open water levels are constant and there is full contact between the surface water

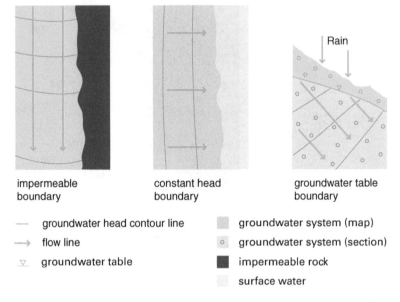

		Rain
impermeable boundary	constant head boundary	groundwater table boundary

Figure 3.16. Two maps and
a cross section (on the right)
indicating the hydraulic head
contour lines and flow lines at
boundaries.

— groundwater head contour line

→ flow line

▽ groundwater table

    groundwater system (map)

○   groundwater system (section)

    impermeable rock

    surface water

and the groundwater system, then the boundary can be considered as a hydraulic head contour line. The flow lines are perpendicular to this contour line and the open water boundary.

– *Groundwater table boundary*: This boundary is the upper boundary of a groundwater system that may be influenced by recharge or discharge or none of these phenomena at all (see section 1.2.5). First, consider the case of recharge or discharge. The flow lines and the hydraulic head contours lines are both at an angle to the boundary. In case there is no recharge or discharge then the groundwater table acts as an impermeable boundary. The flow lines are parallel to the table and the contour lines of hydraulic head are perpendicular.

How can one use flownets for the calculation of groundwater flow? To answer this question one will have to take a closer look at the flow lines and the pattern of rectangles or squares that makes up a flownet. The flow lines are in fact the limits of so-called 'stream tubes'. Groundwater which is flowing through an individual stream tube is not loosing, neither gaining groundwater from the neighbouring tubes. Based on this characteristic of a stream tube, the flow rate through an aquifer can be calculated. The computation of the groundwater flow rate in this way is more precise than to estimate its value from 'simple computations' (see above). This is in particular true for complex groundwater flow patterns where the contour lines are not straight and parallel.

Darcy's Law can be engaged to derive expressions for flownet computations. Imagine a flownet for an aquifer and the flow through an individual stream tube at a selected cross section. The groundwater flow rate through the tube at the selected cross section can be computed by taking into consideration the discrete difference in hydraulic head for the two nearest contour lines, the width of the stream tube and the thickness of the aquifer. Taking as a starting point the general Darcy's Law equation (equation 3.2), then the following expression is obtained:

$$Q_s = -K\,H\,w_s\,\frac{\Delta\phi}{\Delta s} \tag{3.48}$$

where:
Q_s = flow rate through the stream tube (m^3/day)
K = horizontal coefficient of permeability (m/day)
H = saturated aquifer thickness (m)
w_s = width of the stream tube (m)
$\Delta\phi$ = discrete difference in hydraulic head (m)
Δs = discrete distance between contour lines (m)

The flow rate through the whole aquifer at the selected cross section can then be calculated by summation of the flow rates through the individual stream tubes. In case the subscripts 1, 2, etc. are introduced to refer to the first, second, etc. stream tube, then the following expression can be used to compute the (total) flow rate:

$$Q = -[K_1\,H_1\,w_{s1}\,\frac{\Delta\phi}{\Delta s_1} + K_2\,H_2\,w_{s2}\,\frac{\Delta\phi}{\Delta s_2} + \ldots\ldots\ldots] \tag{3.49}$$

Expression (3.49) can be further simplified and evaluated by considering 'squares'. At the cross section where the flow is computed, the flow lines can be drawn in such a way that the width of a stream tube is equal to the distance between the contour lines. Thus, $w_{s1} = \Delta s_1$, $w_{s2} = \Delta s_2$ etc. Assume that there are a total of n_s stream tubes for the whole width of the aquifer. Then, for the case of constant transmissivity ($T = K_1 H_1 = K_2 H_2$, etc.), the flow rate through the aquifer can be computed with:

$$Q = -n_s T\,\Delta\phi \tag{3.50}$$

Figure 3.17 highlights the procedure for a flownet computation. Groundwater head contour lines as shown on the map can be compiled for the groundwater system consisting of a single aquifer. The flow in the aquifer is directed towards the lake. The flow through (part of) the aquifer can be computed at the shown, curved, cross section. Perpendicular to the contour lines nearest to the cross section, flow

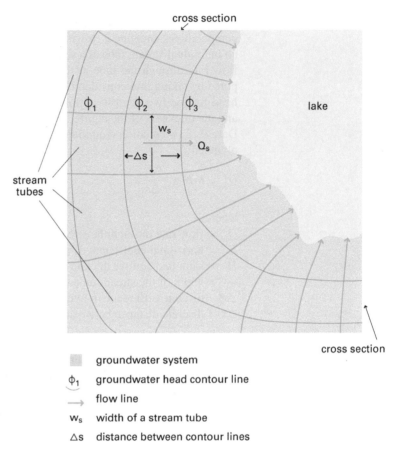

Figure 3.17. Map showing an aquifer bordering a lake area. The drawn flownet can be used to compute the flow rate of groundwater towards the lake.

groundwater system

ϕ_1 groundwater head contour line

⟶   flow line

w_s width of a stream tube

Δs distance between contour lines

lines can be drawn on the map, whereby the various widths of the created stream tubes are kept equal to the distances between the contour lines. Through extension of the flow lines, the full flownet is composed. The discrete difference in hydraulic head can be determined by subtracting the heads of the two contour lines at the cross section. The number of stream tubes can be counted on the map. The transmissivity can best be assessed from the evaluation of pumping tests. In case the transmissivity is constant at the cross section, then equation (3.50) can be engaged to compute the flow rate through the aquifer.

Modelling of regional flow

Regional groundwater flow can be modelled with numerical groundwater models. Before 1980, techniques based on physical models were quite extensively used to simulate regional flow in groundwater systems. In the '80s, numerical models replaced the physical models. In particular the ease of the computer and the increase in their computation speed has given a tremendous boost to the engagement of numerical models. When flow computations on a regional scale have to be carried out, computerised numerical models are superior to calculation methods based on 'simple computations' and flownets. However, these latter techniques remain useful as they give quick estimates on flow rates and provide input for the numerical models.

With numerical groundwater models, groundwater flow rates and hydraulic heads can be computed at the cells of a model grid that covers the entire, or part of a considered groundwater system. Depending on the type of model used, not only the flows and the heads can be computed, but other parameters may be determined as well. The most common types of models are:

i) *Groundwater flow models.* These models concentrate on the computation of groundwater flow rates and hydraulic heads in a groundwater system. The models can be used for the optimisation of coefficients of permeabilities, storativities, recharge rates, or other parameters. Groundwater balances compiled on the basis of the optimised parameters may lead to assessments on the amount of groundwater that is available in a groundwater system (see section 6.3). Groundwater flow models are also used to simulate the effect of groundwater abstractions or other human activities on a groundwater system. Groundwater flow models may also have options to calculate flow lines and travel times. These facilities assist in estimating the behaviour of contaminants in groundwater and help in the design of appropriate protection zones around abstraction wells.

ii) *Fresh-saline interface models.* In addition to groundwater flow rates and hydraulic heads, these models compute the positions of sharp fresh-saline interfaces in groundwater systems. They can be used for similar purposes as outlined for the flow models. At the same time, however, they are able to simulate the effects of human activities, including groundwater abstractions, on the position of the interface.

iii) *Solute transport models.* Based on the so-called 'groundwater flow field' computed with a flow model, these models calculate the concentrations of solutes in a groundwater system. Superior to the groundwater flow models with flow line and travel time computational facilities, the transport models are used to estimate the effects of groundwater contamination and to design protection zones.

iv) *Unsaturated flow models.* These models compute groundwater flow rates and groundwater pressure distributions, not in the – saturated – groundwater system, but in the unsaturated zone. The models are often engaged to assess the effect of human activities on the water distribution in the unsaturated zone and the growth of agricultural crops.

v) *Statistical flow models.* These models are extended with statistical tests. The extension of a flow model with the Kalman filtering technique is a good example. Statistical tests can be used to eliminate model errors, so that model results are improved.

vi) *Groundwater management models.* Rapidly becoming popular, these models are flow models that also have modules to optimise well abstractions in a groundwater system. The models supply information on the best locations for well fields and optimum abstraction rates.

For the modelling of regional flow in a groundwater system, a stepwise procedure is usually followed. The steps to be taken to build a flow model could be defined and described as follows:

– *Drafting of modelling objectives.* The first step concerns the formulation of the modelling objectives to be defined in agreement with general study objectives (see section 6.3). For example, the modelling objective may concern the assessment of the effects of well abstractions on regional groundwater flow rates and hydraulic heads. To achieve this objective, the engagement of a groundwater flow model could be considered.

– *Formulation of the 'conceptual' model.* This step includes a thorough hydrogeological evaluation of the various rock types, which are present in the groundwater system to be modelled. The rocks are schematised into aquifers, aquitards, and aquicludes (see section 2.1.2). Model conceptualisation then includes the representation of these units, in particular the aquifers, as so-called 'model layers'. An aquifer may form an individual model layer, but several aquifers may also be grouped together to form one model layer. Alternatively, one aquifer may be sub-divided into smaller units to form several layers.

The conceptual modelling activities also include the determination of the model area boundaries and the formulation of the groundwater balance for the groundwater system. The boundaries are selected at considerable distances from the locations where human activities are to be simulated with the model. Two types of model boundaries are usually considered. They include the type where the flow is prescribed, referred to as a 'flow' or 'flux' boundary, and the type where the hydraulic head is specified. The latter type of boundary is called a 'specified head' boundary. The activities concerning the groundwater balance include the identification of major recharge and discharge terms, and the rough delineation of model areas where these terms apply.

– *Preparation of model data.* This step includes the preparation of typical model data including 'tops and bottoms' of the model layers, coefficients of permeability, resistances, storativities, recharge and discharge data, etc. These data may be prepared by plotting them on maps, or alternatively, computerised data files can be set up in spreadsheets or more advanced databases.

– *Preparation of the model grid and input of model data.* During this step, the model grid is designed. The design depends on the selected computer code for the modelling. Computer codes are based on typical computational methods, which require a specific grid set up. The grid may consist of cells with the shape of squares, rectangles, triangles or polygons. The model may have a 'coarse grid', but in sub-areas where one wishes to have more detail the grid may also be 'refined'.

Model data input concerns the assignment of data to the model. A transparent sheet of paper showing the grid can be super-imposed on the maps with the prepared groundwater data. Data are read off from the map and are input into the computer to prepare computerised model data files. As an alternative, data files already stored in databases can be converted into model data files. The conversion can be effectuated using interpolation programs that determine the correct parameter values at the appropriate grid cells.

– *Model calibration.* During the model calibration step, computed values for groundwater flow rates and hydraulic heads are compared with values observed in the field. In case the computed and observed values do not match, then the differences between the values are analysed. The differences may be attributed to an erroneous input or incorrect assessment of one or more of the model parameters. Worst of all, however, is a 'wrong' conceptual model. Subsequently, the model is corrected and the computations are done again. This process is repeated until a satisfactory match between computed and observed values is obtained.

– *Model simulation.* During model simulation, various scenarios of human activities to be carried out in the modelled area are tried out. These activities may include the initiation of groundwater abstractions, the installation of artificial recharge schemes, or the implementation of changes in the surface water system. With the model, the effects of these activities on the groundwater system can be predicted. These effects can be computed in terms of altered groundwater flow rates and hydraulic head changes. An optimum scenario will be selected for implementation.

Figures 3.18 and 3.19 illustrate the 'conceptual model' and 'grid design' steps related to a regional modelling study carried out in the northeastern part of The Netherlands. The study was initiated to assess the effects of abstraction wells located at a site named Valtherbos. The area is underlain by a groundwater system consisting of a series of aquifers made up of fine and coarse sands. The sands are alternated with

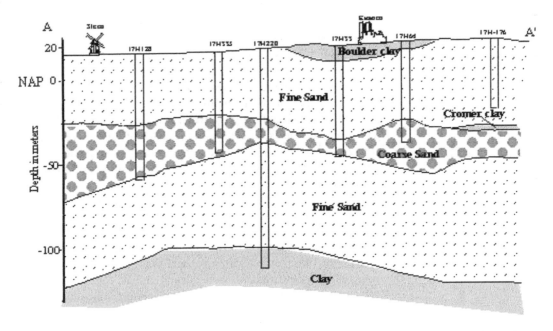

Figure 3.18. Schematic cross section showing the hydrogeology considered for the building of a regional groundwater flow model in the Valtherbos area, Emmen, The Netherlands (Zhang, 1996).

horizons containing clay lenses which can be considered as aquitards. At about 120 m below surface, the base of the groundwater system consists of very compact clays with a negligible coefficient of permeability. During the formulation of the conceptual model, the groundwater system was schematised into four model layers.

Figure 3.18 shows that the model layers can be represented successively by a thin top layer including near surface features like the Boulder clay, and another three layers representing fine sandy, coarse sandy and again fine sandy material. The aquitards including the Cromer clay, were not simulated by separate model layers, but were represented by resistances. Figure 3.19, presenting the model grid for the area, indicates the rectangular nature of the grid cells and the grid refinement at the abstraction wells at Valtherbos. With the model, the effects of the abstractions on the regional groundwater flow rates and hydraulic heads could be computed.

During model calibration and model simulation, calculations are carried out following typical numerical methods. Generally, one distinguishes two computational methods: the 'finite difference' method and the 'finite element' method. In this textbook, only the finite difference method will be elaborated. This method requires a model grid consisting of cells with the form of squares or rectangles. For so-called

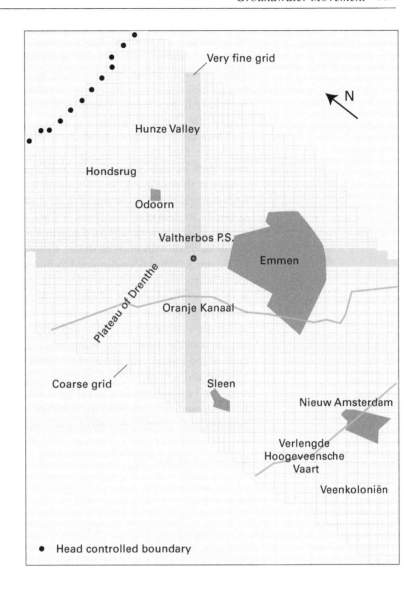

Figure 3.19. Computer printout of map showing the model grid for the model building in the Valtherbos area, Emmen, The Netherlands (Zhang, 1996).

'block centered' finite-difference groundwater models, groundwater flow rates and hydraulic heads are computed at the centres of the grid cells. One refers to 'face centered' models if the computations are carried out at the corners of the grid cells.

Groundwater model computations following the finite difference method are also based on the Darcy's Law and continuity equations. One has to realise that in modelling grids are considered that consist of cells that have discrete dimensions. In other words, the sides of the cells have a finite well-defined length and width, and the grid covers a groundwater system represented by model layers with a finite thickness. Thus, the Darcy's Law and continuity equations in differential form have to be adapted. Instead of using 'differentials', one will have to settle for 'finite-differences'.

There are two options to work out the Darcy's Law and continuity equations. One option is to work out the combined differential equations such as (3.14) and (3.17). This option will not be further discussed in this textbook, but details on the complete derivation of the relevant equations can be found in Bear and Verruijt (1987). The other option is to discretise the Darcy's Law and the continuity equations separately (Olsthoorn, 1985). Since this option is easier to understand, it will be dealt with in more detail, using a simple case.

Figure 3.20 has been designed to illustrate the simple case. The figure presents a model grid covering a groundwater system consisting of a single confined aquifer of constant thickness. The groundwater system is assumed to be in the non-steady state. The conceptual model set up for the system shows one model layer representing the aquifer. The thickness of the model layer is equal to the thickness of the aquifer. The cells of the model grid are considered to be rectangular. The Darcy's Law equations in the form of expression (3.3), (3.4) or (3.5), without the sign, can be discretised on a cell-by-cell basis. The resulting finite-difference equation expressing the horizontal flow rate in the model layer between, for example, the cells 5 and 1, can be formulated as:

$$Q_{1,5} = K_{1,5} \, H \, \Delta x \, \frac{\phi_1 - \phi_5}{\Delta y_1} \tag{3.51}$$

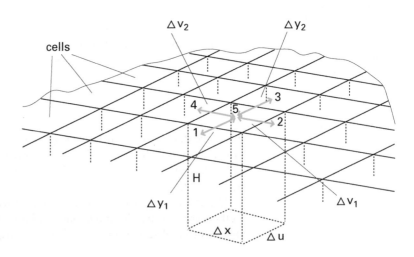

Figure 3.20. Model grid covering a single confined aquifer of constant thickness.

where:

$Q_{1,5}$ = horizontal flow rate in the model layer between cell 5 and 1 (m^3/day)

$K_{1,5}$ = horizontal coefficient of permeability at cell 5 and 1 (m/day)

H = aquifer thickness (m)

Δx = width of cell 5 (m)

Δy_1 = distance between the centres of cell 5 and 1 (m)

ϕ_5; ϕ_1 = hydraulic heads in the model layer at the centres of cell 5 and 1 (m)

The continuity equation represented by equation (3.15) can also be discretised. The finite-difference representation of the first two terms of the left side of this equation is equal to the sum of the horizontal flow rates in the model layer between cell 5 and the cells 1, 2, 3 and 4. Flows towards the centre of cell 5 are considered positive. Realising that groundwater storage is not related to a control volume of 1 m^3, but to the volume of an aquifer column with a cross sectional area equal to the surface area of cell 5, the finite-difference form of the right side of equation (3.15) can also be considered. The right hand side is the product of the storativity and the surface area of cell 5, and the discrete difference in hydraulic head for a selected time step. The complete finite-difference equation can be approximated as follows:

$$Q_{1,5} + Q_{2,5} + Q_{3,5} + Q_{4,5} = S \, \Delta x \Delta u \, \frac{[\phi_{5,t=t} - \phi_{5,t=t-\Delta t}]}{\Delta t} \qquad (3.52)$$

where:

S = storativity of the model layer at cell 5 (dimensionless)

Δu = length of cell 5 (m)

$\phi_{5,t=t}$ = hydraulic head in the model layer at cell 5, at end of time step (m)

$\phi_{5,t=t-\Delta t}$ = hydraulic head in the model layer at cell 5, at start of time step (m)

Δt = time step length (days)

The Darcy's Law equation and the continuity equation in finite-difference form (equations 3.51 and 3.52) can be combined with each other. For the other horizontal flows $Q_{2,5}$, $Q_{3,5}$, $Q_{4,5}$ in the model layer between cell 5 and cells 2, 3, and 4, expressions similar to equation (3.51) can be found. Inserting the equations into expression (3.52) yields:

$$K_{1,5}\,H\Delta x\,\frac{\phi_1 - \phi_5}{\Delta y_1} + K_{2,5}\,H\Delta u\,\frac{\phi_2 - \phi_5}{\Delta v_1} + K_{3,5}H\Delta x\,\frac{\phi_3 - \phi_5}{\Delta y_2} + K_{4,5}H\Delta u\,\frac{\phi_4 - \phi_5}{\Delta v_2}$$

$$= S\Delta x\Delta u\,\frac{[\phi_{5,t=t} - \phi_{5,t=t-\Delta t}]}{\Delta t} \tag{3.53}$$

For the simple case considered, equation (3.53) can be looked upon as a composite finite-difference equation for groundwater flow. One could add to this numerical equation so-called 'source or sink' terms. These terms indicate the exchange of water with the 'outside world'. A sink term could represent groundwater abstractions from the aquifer model layer at cell 5. Alternatively, a source term could indicate water added to the model layer by an infiltration gallery or injection well as part of an artificial recharge scheme. With the source or sink term included, equation (3.53) can be written as:

$$K_{1,5}\,H\Delta x\,\frac{\phi_1 - \phi_5}{\Delta y_1} + K_{2,5}\,H\Delta u\,\frac{\phi_2 - \phi_5}{\Delta v_1} + K_{3,5}H\Delta x\,\frac{\phi_3 - \phi_5}{\Delta y_2} +$$

$$K_{4,5}H\Delta u\,\frac{\phi_4 - \phi_5}{\Delta v_2} + W = S\Delta x\Delta u\,\frac{[\phi_{5,t=t} - \phi_{5,t=t-\Delta t}]}{\Delta t} \tag{3.54}$$

where:
W = source or sink term (m^3/day)

Composite finite-difference equations can be solved for the computation of groundwater flow rates and hydraulic heads. The heads are usually computed first and then the flow rates are calculated. Two types of procedures can be engaged to solve the heads in the equations: 'implicit and explicit solution procedures'. In the implicit procedures, the hydraulic head is solved implicitly; in explicit procedures the head is computed in an explicit form. By insertion of the calculated hydraulic heads in the Darcy's Law equations in finite-difference form, the flow rates can be computed.

For the simple case, the explicit solution procedure will be elaborated. To avoid very lengthy formulas, the case is simplified. Assume that the model grid consists of squares: $\Delta x = \Delta y_1 = \Delta u = \Delta v_1$, etc. Imagine also that the aquifer is homogeneous and isotropic: $K = K_{1,5} = K_{2,5} = K_{3,5}$, etc. Instead of the non-steady state, the steady state condition is considered. Finally, the sink or source term is also neglected. Equation (3.54) can then be written as follows:

$$KH\,[\phi_1 + \phi_2 + \phi_3 + \phi_4 - 4\phi_5] = 0 \tag{3.55}$$

By re-arranging the terms of equation (3.55), an explicit expression for the hydraulic head can be obtained. This head, representative for the model layer at the centre of cell 5, is expressed by the following equation:

$$\phi_5 = \frac{[\phi_1 + \phi_2 + \phi_3 + \phi_4]}{4} \tag{3.56}$$

Equation (3.56) indicates that the hydraulic head in the model layer at cell 5 is simply the average of the heads at the surrounding cells. Flow rates can be computed using the hydraulic head obtained for the model layer at cell 5. The obtained value can be inserted into finite-difference expressions of Darcy's Law in the form of equation (3.51). The horizontal groundwater flow rates in the model layer between cell 5 and cells 1, 2, 3 and 4 can then be computed. By summation of the flow rates, the total horizontal inflow or outflow rates for the model layer at cell 5 can be assessed.

Using a simple case, the computations of hydraulic heads and flow rates for a single cell have been explained above. The question arises how the computations are organised for the complete model grid? The procedure will not be outlined for the simple case, but will be discussed in a general sense. A so-called 'iteration procedure' organises the computations for all the cells in the grid. The iteration procedure takes care of the computation of the hydraulic heads and flow rates on a cell by cell basis, usually in a row- or column-wise fashion. Initial hydraulic heads and at least one constant head are needed to perform the iteration procedure in an adequate manner. An iteration is said to be completed when one round of computations has been done for all the cells in the model. Successive iterations are carried out until the computed hydraulic heads and flow rates hardly change anymore and the computations are terminated by pre-set criteria.

The model calculations are carried out by computer codes, which are able to substantially reduce the time needed for completing the iterations. In recent years many computer codes for groundwater modelling have been developed. For example, explicit solution techniques have been used to design groundwater models based on spreadsheets including Lotus 123 and Excel. Groundwater modelling codes using implicit procedures are, for example, MODFLOW and MICROFEM. MOD-FLOW is the quasi- and fully three-dimensional code developed by the United States Geological Survey (McDonald & Harbaugh, 1988). Codes for the calculation of travel time, for transport, for parameter optimisation and well field planning can easily be coupled to MODFLOW. The MICROFEM program developed in The Netherlands is based on the finite element method (Hemker & Van Elburg, 1987).

3.2.3 *Local groundwater flow*

The concept of local flow   Regional groundwater flow in groundwater systems usually covers large areas. This flow may be influenced by activities of a local nature. For example, the construction of canals may influence the regional flow of groundwater. Another example concerns wells. Abstractions from wells may also affect the regional flow of groundwater. These local influences on regional flow can often be observed on groundwater head contour maps. For the case of canals, contour lines for the hydraulic heads which are parallel or skewed to the canals may be shown and around wells or well fields, the contour lines on the maps may follow a circular or ellipsoidal pattern.

Groundwater flow associated with local activities will be referred to as 'local groundwater flow'. Computations concerning local groundwater flow can also be done. Estimates on the local flow of groundwater can be obtained using methods based on 'simple computations', flownets or numerical groundwater models (see section 3.2.2). As a tradition, however, these local groundwater flow problems are solved by analytical methods (Verruijt, 1970). These methods will only briefly be addressed in this textbook. Basic cases of the flow between canals and the flow to a well will be presented.

Flow between canals   Consider the case of the local flow of groundwater between canals. What are the expressions describing the hydraulic heads and the flow rates of groundwater between canals? Figure 3.21 shows the case that

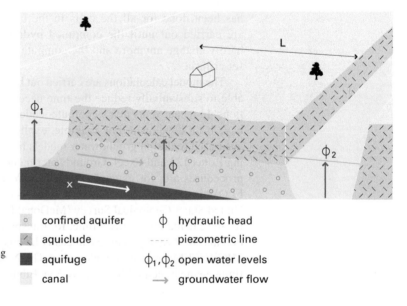

Figure 3.21. Diagram showing an aquifer with flow between canals.

∘ confined aquifer	ϕ hydraulic head
aquiclude	---- piezometric line
aquifuge	ϕ_1, ϕ_2 open water levels
canal	→ groundwater flow

one may have in mind: a groundwater system consisting of a confined aquifer of sandy material with constant thickness and permeability, overlain by an aquiclude of impermeable clays. The aquifer boundaries are two canals at a distance L apart. They have constant open water levels ϕ_1 and ϕ_2. The flow is from the left canal to the right canal and is perpendicular to these canals. The flow can be assumed horizontal and one-dimensional; say in the x direction. The flow is also considered to be in a steady state. The relevant Darcy's Law equation (3.3) is as follows:

$$q_x = -K_x \frac{\partial \phi}{\partial x} \tag{3.57}$$

For one-dimensional flow in the x direction the continuity equation (3.14) for steady flow simplifies to:

$$\frac{\partial^2 \phi}{\partial x^2} = 0 \tag{3.58}$$

To find a useful expression for the hydraulic head, equation (3.58) can be integrated:

$$\frac{\partial \phi}{\partial x} = C_1 \tag{3.59}$$

Then, integrating expression (3.59) yields:

$$\phi = C_1 x + C_2 \tag{3.60}$$

Expression (3.60) shows the integration constants C_1 and C_2 that have to be solved from boundary conditions. To find C_2, one can use the following condition. At $x = 0$, the hydraulic head equals the open water level of the left canal: $\phi = \phi_1$ (see Figure 3.21). Insertion into expression (3.60) yields:

$$C_2 = \phi_1 \tag{3.61}$$

To determine C_1 another condition can be considered. At $x = L$ the hydraulic head is the same as the open water level of the canal on the right: $\phi = \phi_2$. Combination of this condition with equations (3.60) and (3.61) yields:

$$C_1 = \frac{\phi_2 - \phi_1}{L} \tag{3.62}$$

The integration constants can be inserted into equation (3.60). Then the following expression is obtained for the hydraulic head:

$$\phi = \frac{\phi_2 - \phi_1}{L} x + \phi_1 \tag{3.63}$$

Equation (3.63) shows that the hydraulic head in the aquifer is a function of the distance x from the (left) canal. The gradient of the hydraulic head is constant which means that there is a linear relationship between the hydraulic head in the aquifer and the distance (x).

The groundwater flow rate through a unit area of 1 m² can be obtained by differentiating equation (3.63) and combining the result with the relevant Darcy's Law equation (3.3). The following expression is obtained:

$$q_x = -K_x \frac{\phi_2 - \phi_1}{L} \tag{3.64}$$

From equation (3.64) it can be deduced that the flow rate through a unit area of 1 m² in the aquifer can be considered constant. This result could have been expected for a confined aquifer of constant thickness and permeability, and steady state groundwater flow.

Flow to a well

One may also try to find expressions for the hydraulic head and the flow rate of groundwater near an abstraction well. Take the case portrayed in Figure 3.22 and note, in Figure 3.23, the pictural view of a groundwater

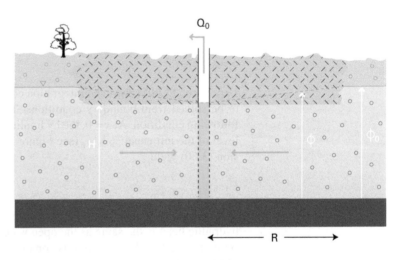

Figure 3.22. Cross section through an aquifer showing the flow towards an abstraction well.

○	aquifer	ϕ	hydraulic head
	aquiclude	ϕ_0	constant hydraulic head
	aquifuge	Q_0	abstraction rate
→	groundwater flow	H	aquifer thickness
▽	groundwater table	----	piezometric line

abstraction well in the Rada Basin in Yemen. Figure 3.22 shows a well in a groundwater system consisting of a mainly confined aquifer. The aquifer consists of sandy material of constant permeability and is over- and underlain by clayey aquicludes and aquifuges. The aquifer is extensive and at a large distance R from the well, a constant hydraulic head, ϕ_0, will be assumed. At distances larger than R, the aquifer is unconfined. In this part of the aquifer, the influence of the abstraction well on the hydraulic head is small, as a result of recharge. Imagine that the flow in the aquifer is horizontal and is radially directed towards the well. Finally, assume that the flow is in a steady state and that the constant abstraction rate, Q_0, at the well is known.

For the derivation of the relevant expressions, Darcy's Law equation for radial flow, which is similar to equation (3.3), (3.4) or (3.5), can be considered:

$$q_r = -K_r \frac{d\phi}{dr} \tag{3.65}$$

where:
q_r = radial flow rate through a unit area of 1 m^2 (m/day)
K_r = coefficient of permeability in a radial direction (m/day)
r = distance from the abstraction well (m)

Figure 3.23. Groundwater pumped from a deep abstraction well in the Rada Basin, Yemen.

The continuity equation for steady radial flow can be obtained from continuity equation (3.14), defined for flow in the x, y and z directions. To find an appropriate expression for the radial flow, one will first have to reduce equation (3.14) to represent horizontal flow:

$$\frac{\partial^2 \phi}{\partial x^2} + \frac{\partial^2 \phi}{\partial y^2} = 0 \tag{3.66}$$

The conversion of the continuity equation (3.66) for horizontal flow in the x and y direction to the continuity equation for radial flow will not be elaborated in this textbook. The result, however, is:

$$\frac{d^2 \phi}{dr^2} + \frac{1}{r}\frac{d\phi}{dr} = 0 \tag{3.67}$$

There are several ways to solve equations (3.65) and (3.67) in terms of hydraulic heads and groundwater flow rates. The simplest approach is to take the Darcy's Law equation as a starting point. One will find in this approach that the continuity equation (3.67) will not be applied directly, but only the continuity concept will be considered. The full approach is as follows. Instead of taking the Darcy's Law equation for the radial flow rate per unit area of 1 m^2 (equation 3.65), the total flow rate in the radial direction will be considered. This flow rate can be represented by the flow rate through a circular ring at a distance, r, from the well. The ring has a height equal to the thickness of the aquifer. Equation (3.65) can then be elaborated into:

$$Q_r = -2\pi r H K_r \frac{d\phi}{dr} \tag{3.68}$$

where:
Q_r = total radial flow rate towards the well (m^3/day)
H = thickness of the aquifer (m)

The continuity concept can be visualised as follows. In the aquifer, there are no 'water gains' or 'water losses' in the z direction, and since steady flow is considered, there is also no change in groundwater storage. This means that the total radial flow rate is constant and equal to the (constant) abstraction rate at the well: $Q_r = -Q_o$. This relationship can be incorporated into equation (3.68) as follows:

$$\frac{d\phi}{dr} = \frac{Q_o}{2\pi K_r H}\frac{1}{r} \tag{3.69}$$

Expression (3.69) can be integrated whereby the following result is obtained:

$$\phi = \frac{Q_o}{2\pi K_r H} \ln r + C \qquad (3.70)$$

The C has to be solved from boundary conditions. At the large distance, $r = R$, from the abstraction well, the hydraulic head is constant: $\phi = \phi_0$. When the above condition is inserted into equation (3.70) then the following integration constant is obtained:

$$C = \phi_0 - \frac{Q_o}{2\pi K_r H} \ln R \qquad (3.71)$$

Inserting the integration constant into equation (3.70), leads to the following expression for the hydraulic head:

$$\phi = \phi_0 + \frac{Q_o}{2\pi K_r H} \ln \frac{r}{R} \qquad (3.72)$$

Expression (3.72) indicates the logarithmic relationship between the hydraulic head and the distance to the well. The cross-sectional shape of the cone of depression that develops as a result of abstracting water from the well is described by this expression.

The radial flow rate of groundwater through a unit area of 1 m^2 can be deduced by combining expression (3.65) with (3.69):

$$q_r = -\frac{Q_o}{2\pi H} \frac{1}{r} \qquad (3.73)$$

Equation (3.73) shows the linear relation between the flow rate per unit area and the distance from the well. The larger the distance from the well is taken, the smaller is the flow rate per unit area. Note that for $r = 0$, expression (3.70) is not valid.

3.2.4 *Pumping test methodology*

Field technique

Formulae describing the local groundwater flow at a well are engaged to interpret the results of pumping tests which, for example, are conducted upon completion of exploration drilling (see section 2.3.2). Pumping tests can be carried out after a drilled hole has been completed into a well. Essentially, pumping tests are tests whereby the well is pumped for a considerable period while the effect on groundwater levels (hydraulic heads) is measured. After a pump has been lowered in the well and has been switched on, groundwater levels in the pumped well and in nearby observation wells are recorded. In addition, abstraction rates are assessed and water samples are also taken during the execution of a pumping test.

Figure 3.24 presents a team carrying out a pumping test in an existing well drilled in the sandy Tihama Plain in Yemen. The groundwater

Figure 3.24. Pumping test carried out at an agricultural well in the Tihama Plain, Yemen.

levels in the pumped well were recorded with a manual gauge whereby a light switches on when the water is reached. The abstraction rates were assessed outside the pump house, using an empty oil barrel.

Interpretation method   Before a pumping test can be interpreted, the measured groundwater levels are usually corrected for effects other than the pumping in the well. These effects may relate to the occurrence of recharge from precipitation and groundwater discharge during the test, to adjustments of open water levels, or to oceanic or moon-driven tides. The interpretation of the pumping test itself usually starts with the assessment of the local geo-hydrological condition of the groundwater system (Kruseman & De Ridder, 1990). A confined, semi-confined, semi-unconfined or unconfined condition is judged from the information obtained from rock sampling and geophysical logging, and from the shape of a plot of measured groundwater levels against time. Using well formulae, the interpretation of the values of the transmissivities, vertical resistances and storativities of the groundwater system is then carried out.

Many of the so-called 'constant yield' tests can be analysed with personal computer programs. Figure 3.25 shows the computer analysis

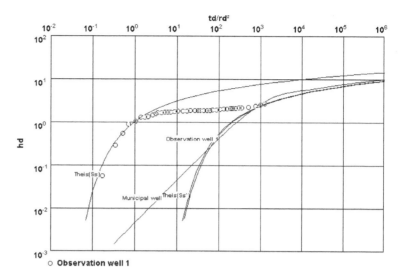

Figure 3.25. Computer print-out of an interpretation of a pumping test in the Kenhardt area, South Africa.

of a test carried out in the groundwater system in the Kenhardt area in South Africa (Ministry of Water Affairs, 1979). The system, introduced in section 1.2.5, consists of fractured metamorphic rock overlain by confining clayey weathered rock. Several parameters typical for fractured rock were successfully evaluated with a computer code. The transmissivity of the fractured aquifer is in the order of 120 m²/day, the horizontal coefficient of permeability is about 9.5 m/day and the total storativity (fractures and block matrix) amounts to about 0.0005.

3.3 THE FLOW SYSTEMS CONCEPT

3.3.1 *Formation of flow systems*

The most common ways to describe the flow in a groundwater system and to carry out flow computations are discussed in section 3.2. The basis for the descriptions and computations is the hydrogeological classification of rock types in a groundwater system into aquifers and aquitards. Groundwater flows are computed for these units whereby the assumption of horizontal flow in the aquifers and vertical flow in the aquitards is usually applied.

Another approach to describe the flow in a groundwater system is based on the flow systems concept. This concept relates to the recharge and discharge mechanisms in a groundwater system. In sections 1.2 and 4.1, recharge from precipitation, streams and rivers, and the discharge by springs and capillary flow are extensively discussed. Recharge will finally emerge as discharge. A flow system can be defined as 'a recharge-discharge unit in a groundwater system'. One can imagine that in a groundwater system, there are several flow systems each consisting of representative recharge and discharge units.

Flow system terminology Figure 3.26 illustrates the concept and related terminology of flow systems. The figure presents, schematically, a vertical cross section that shows part of a groundwater system consisting of permeable rock underlain by an impermeable base. The rock is homogeneous and isotropic. In the areas with a higher topography, there is recharge into the permeable rock. In Figure 3.26, recharge is indicated at section (ed). In areas with a depressed topography, there is groundwater discharge, for example at section (ae). The grey lines are contour lines for the hydraulic heads that are drawn perpendicular to flow lines. In principle, the flow unit covered by the area (abcd) can be looked upon as a flow system.

Specific terminology used in the flow systems approach includes the 'hinge line', and the 'groundwater divide'. The hinge line can be defined as 'the line separating the recharge and discharge areas in a flow system'. In Figure 3.26, point (e) falls on a hinge line. A groundwater divide can be defined as a 'no flow boundary of a flow system'. As the name indicates, no groundwater flow crosses a groundwater divide. In Figure 3.26 the vertical (cd) coincides with a groundwater divide.

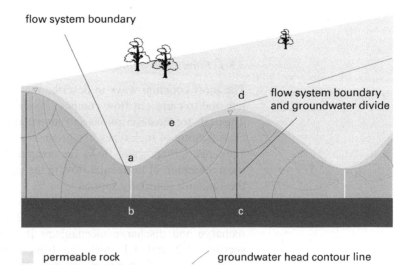

flow system boundary

flow system boundary and groundwater divide

permeable rock

impermeable base

groundwater table

groundwater head contour line

→ flow line

Figure 3.26. Schematic cross section showing the concept of flow systems.

Flow system properties

The topography, geomorphology, geology and geochemistry are responsible for a number of specific properties of flow systems. These properties were first recognised by Toth (1962) when he introduced his views on flow systems. A selection of the properties of a flow system is as follows:

i) In the topographically high recharge areas of flow systems, groundwater flow has substantial downward components. In between recharge and discharge areas, groundwater flow is usually horizontal or sub-horizontal over large distances. This may be especially true for the deeper parts of a groundwater system. In the low discharge areas, there are essentially upward flow components.

ii) The flow directions as described above imply that in the unsaturated zones in the recharge areas the net flow is downward. The flow is upward in the unsaturated zones in the discharge areas.

iii) Groundwater flow can be slow near groundwater divides. In Figure 3.26, for example, the flow of groundwater is slow at the bottom of the groundwater divide (cd).

iv) There may be several flow systems nested into each other (see section 3.3.2).

v) There is usually an increase in total dissolved solids (TDS) concentration of the groundwater during its passage through the flow system. This means that in recharge areas the TDS concentration is relatively low and in discharge zones the concentration is relatively high.

vi) The groundwater in a flow system has its own typical chemical composition which, however, may change in the course of time. The groundwater contains chemical constituents that are related to the chemical composition of its recharge water and the type of rock present in the groundwater system (see section 5.1.2). Groundwater in other flow systems may be quite different in chemical composition.

The properties that are outlined above play an important role in the mapping and identification of flow systems in a groundwater system. Some of the tools that can be used to identify flow systems are:

– *Topographical maps.* These maps show the higher and lower areas of a groundwater system. Areas with a higher topography, relating to recharge areas, and areas with a lower topography, pointing to discharge areas, indicate the distribution of flow systems. The discharge areas may also be represented on the map by abundant vegetation, marshy areas, springs, and surface watercourses.

– *Groundwater head contour maps.* Ideal for the identification of flow systems are contour maps showing hydraulic heads. The contour maps are preferably drawn on the basis of groundwater level measurements taken in observation wells with multiple screens installed at different depths below land surface. Differences in hydraulic head indicate groundwater flow directions that lead to the direct identification of flow systems.

– *Groundwater chemistry maps*. Maps showing TDS concentrations or chemical constituents dissolved in groundwater can also contribute to the identification of flow systems. In addition to the most common inorganic constituents, isotopes like ^{2}H, ^{3}H, ^{14}C and ^{18}O can be analysed in water samples. Isotope concentrations are a measure for groundwater travel times and may lead to assessments on the origin of groundwater. This will help to delineate groundwater recharge areas. Groundwater chemistry maps can be used in combination with other maps, for instance geological maps, to delineate the complete set of recharge and discharge areas.

3.3.2   Flow computations

Flownets and flow systems

What is the approach to the computation of groundwater flow when one considers the flow systems concept? For these systems, one may carry out three-dimensional flow computations. Flow cases, however, are frequently simplified to cases whereby flows are computed in the two-dimensional vertical plane. This simplification assumes that the flow directions at any point in the vertical plane are parallel to this plane. In other words, flow that is within the plane should not 'leave' the plane.

Groundwater flow in the flow system concept is preferably computed using analytical methods or groundwater models. Toth (1962, 1963) followed an analytical approach to compute head distributions and flow rates of groundwater in the vertical plane. The derivation of his equations will not be discussed in this textbook. It should suffice to say that as a starting point Toth formulated the appropriate Darcy's Law and continuity equations. An assumed shape of the groundwater table he used as a boundary condition for solving the Laplace continuity equation. Toth assumed that groundwater tables could be considered as straight lines or lines with a sinusoidal shape. The equations he found were used to compute the hydraulic heads in the vertical plane. Finally, Toth could prepare flownets from the computed hydraulic head distributions, and, he used Dary's Law for the computation of groundwater flow rates.

The analytical approach of Toth had some setbacks. The approach was based on a homogeneous and isotropic groundwater system, which is usually not the case in nature. A disadvantage was also that an idealised shape of the groundwater table had to be assumed. Nowadays, numerical methods incorporated in modern groundwater models are engaged to carry out the flow computations for flow systems. Most of the basic principles of groundwater modelling as outlined in section 3.2

can also be applied to flow system computations. With appropriate models hydraulic heads and flow lines, flow rates of groundwater, and groundwater travel times can be calculated. Presentation of the results is usually in the form of a flownet for the vertical plane. Well-known computer codes, which can carry out the calculations, include FLOWNET (Van Elburg et al., 1989) and MODFLOW in combination with the program MODPATH (Pollock, 1989).

Figure 3.27 illustrates a cross section showing a flownet prepared on the basis of flow computations for an extensive groundwater system in the eastern part of The Netherlands. The system is mainly made up of coarse sands, which are underlain by finer material including sandy silt and clays. Using a prescribed groundwater table, FLOWNET computed the distribution of hydraulic head contour lines and flowlines in the groundwater system. The hydraulic heads were also converted into groundwater flow rates. The flow lines and the indicated flow directions show that there are two main flow systems, discharging into the Oude IJssel river. In addition, one may conclude that the flow mainly

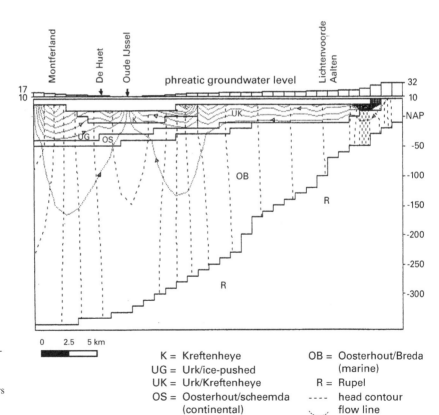

Figure 3.27. Computer printout of a cross section showing flow systems in a sandy groundwater system, Doetinchem area, The Netherlands. Note that the dashed hydraulic head contour lines are not strictly perpendicular to the flow lines. This is due to anisotropy in the sandy layers and scale-distortion (Geochem, 1990).

K = Kreftenheye
UG = Urk/ice-pushed
UK = Urk/Kreftenheye
OS = Oosterhout/scheemda (continental)

OB = Oosterhout/Breda (marine)
R = Rupel
---- head contour
⌣⌣ flow line

takes place in the coarse sands, which are present in the upper 50 m below land surface.

Effect of topography and geology

Flownets can be used to study the effect of topography and geology on a flow system. Figure 3.28 shows two cross sections, composed on the basis of flownets, to illustrate these effects on a system. The upper section presents a groundwater system consisting of homogeneous and permeable rock. The topography has an overall slope from the right to the left, but there are also hills and depressions. This topography leads to the formation of a 'nested flow system'. There are several small flow systems that develop within a larger, more regional system.

The lower cross section represents a groundwater system consisting of two rock types with different permeabilities. The upper rock has a much lower permeability than the rock at the bottom. This difference in permeability has resulted in an almost vertical flow in the upper rock, whereas the flow in the lower rock is mainly horizontal. The case relates to the concept of horizontal flow in the aquifers and vertical flows in the aquitards, as outlined in section 3.2. The section shows the minuscule flow systems that may form in the less permeable rock and the main flow system that develops in the permeable rock.

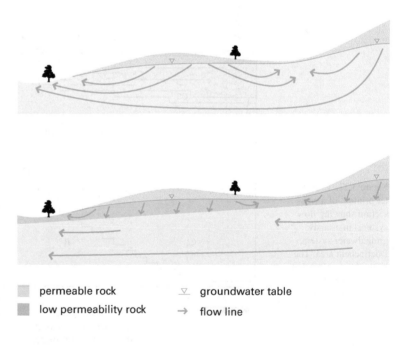

Figure 3.28. The effect of topography (above) and geology (below) on flow systems.

▨ permeable rock	▽ groundwater table
▨ low permeability rock	→ flow line

CHAPTER 4

Water Balances

4.1 GROUNDWATER BALANCE CONCEPTS

4.1.1 *The general idea*

Inventory of components   One can formulate water balances for continents and oceans, for unsaturated zones, or for (saturated) groundwater systems. Most important within the framework of this textbook are the water balances for groundwater systems. These 'groundwater balances' describe the inflows and outflows of groundwater into and out of groundwater systems or parts of these systems. Groundwater stored in or released from these systems is also part of a groundwater balance. Good knowledge of, and insight into the water balance of a groundwater system is of tremendous help when flow computations have to be carried out and the scope for groundwater development has to be assessed.

Groundwater balances can be set up when groundwater recharge and discharge mechanisms of a groundwater system are known. Recharge and discharge are the most essential inflows and outflows of a system and they were introduced and extensively discussed in section 1.2.4.

Figure 4.1 has been prepared to obtain an impression of a groundwater balance and its inflow and outflow terms, using an illustrative case. The cross section in the figure shows part of a groundwater system consisting of an unconfined sandy aquifer, which is underlain by a clayey aquitard and a sandy semi-confined aquifer. The boundaries of the cross-sectional area for which the groundwater balance terms are compiled, are the groundwater table as the upper boundary, the top of the aquitard as the bottom boundary and the left and right edges of the cross section as the vertical boundaries. The boundaries of the selected 'groundwater balance area' may not be the most practical ones (see also below), but they serve well to illustrate the inflow and outflow terms of a water balance.

The inflows into the groundwater balance area consisting of recharge terms and 'other subsurface terms' can be described, and assigned symbols, as follows:

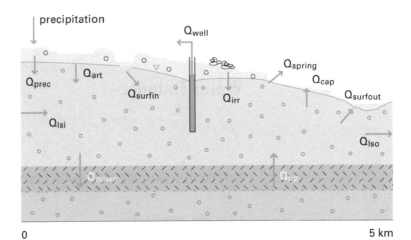

Figure 4.1. Cross section presenting an illustrative case showing the inflows and outflows for setting up a groundwater balance.

- Recharge from precipitation (Q_{prec}). This term describes the inflow of water resulting from surplus precipitation.
- Losses from irrigation (Q_{irr}). This term comprises water losses and leaching requirements, recharging the groundwater where irrigation takes place.
- Inflow from surface water (Q_{surfin}). This recharge term covers the inflow originating from surface water including lakes, streams and rivers.
- Artificial recharge (Q_{art}). This term outlines the inflow of water as engineered by man. This inflow may take place through a system of infiltration ponds, galleries or deep injection wells.
- Lateral subsurface inflow (Q_{lsi}). This term defines the subsurface inflow through the vertical boundaries.
- Vertical subsurface inflow (Q_{up}). This term comprises the inflow through the bottom boundary.

The outflows of the groundwater balance area including discharge terms and 'other subsurface terms' can be summarised as follows:

- Discharge (flow) at springs (Q_{spring}). This term covers the water leaving the system at springs.
- Capillary flow from groundwater tables (Q_{cap}). This discharge term is the upward flow from the groundwater table stimulated by capillary action.
- Outflow to surface water ($Q_{surfout}$). This term refers to the outflow of water into seas, lakes, streams and rivers.

– Abstractions (Q_{well}). This discharge term describes the recovery of groundwater by wells and well fields.
– Lateral subsurface outflow (Q_{lso}). This term defines the subsurface outflow through the vertical boundaries.
– Vertical subsurface outflow (Q_{down}). This term describes the outflow through the bottom boundary.

Most of the groundwater inflow and outflow terms present in natural groundwater systems are considered in the illustrative case shown in Figure 4.1. Perhaps inflow components including recharge in the form of losses from sewer systems and other contaminant sources, and outflows comprising discharge at 'seepage zones' and water captured at subsurface galleries could be added. The above 'list' of inflows and outflows could be used as a sort of 'checklist' for the identification of these terms in a particular groundwater system. The use of a 'checklist' should lead to the assessment of a proper set of groundwater balance terms.

Formulation of the groundwater balance

A groundwater balance describes inflows and outflows for a groundwater balance area. What is the relation between these terms? Generally speaking, for a groundwater balance, the difference between the total inflows and outflows is equal to any changes in groundwater storage. A groundwater balance is always compiled for a pre-determined observation period. In case changes in groundwater storage are indeed assessed during this period, the groundwater balance is referred to as a 'non-equilibrium groundwater balance'. The following relationship can be set up for a non-equilibrium balance:

$$I - O = S_{bal} \tag{4.1}$$

where:
I = total rate of groundwater inflow (m³/day)
O = total rate of groundwater outflow (m³/day)
S_{bal} = volumetric rate of water stored or released (m³/day)

The I and O terms in equation (4.1) can be broken down into the individual inflows and outflows of the groundwater balance. For example, the terms identified for the illustrative case presented in Figure 4.1 could be considered. The worked out groundwater balance, using the symbols as defined in the previous section, is as follows:

$$[Q_{prec} + Q_{irr} + Q_{surfin} + Q_{art} + Q_{lsi} + Q_{up}] - [Q_{spring} + Q_{cap}$$

$$+ Q_{surfout} + Q_{well} + Q_{lso} + Q_{down}] = S_{bal} \tag{4.2}$$

Note that the individual inflow, outflow, and storage terms in equation (4.2) are representing rates, which can be expressed in m³/day. Also note that poor estimates of the inflow, outflow and storage terms may lead to an inaccurate groundwater balance.

Groundwater balances include inflow and outflow terms, some of which may be very difficult to determine. One may select the boundaries of a groundwater balance area in a practical way in order to reduce the number of inflow and outflow terms. This means that it is handy to take the groundwater table as the upper boundary of the groundwater balance area. Trying to include the unsaturated zone above the groundwater table may lead to more terms and complications. The bottom boundary of the groundwater balance area is preferably the top of the aquifuge underlying the groundwater system. Lateral boundaries could favourably be selected along surface water features including lakes, streams and rivers, or along contacts with impermeable aquifuges.

Figure 4.2 presents an illustrative example to demonstrate a practical selection of the boundaries for a groundwater balance area. The cross section in the figure shows a groundwater system that is embedded in a river valley. The system consists of a sandy alluvial aquifer underlain by an impermeable aquifuge that is made up of intrusive granites. Preferably, one selects the boundaries of the groundwater balance area along the shown groundwater table, along the contacts with the impermeable granites, and below the river. For this area the groundwater balance consists of a set of inflow and outflow terms which are indicated in Figure 4.2 by symbols and arrows. The inflows include recharge from precipitation (Q_{prec}), inflow from surface water through a canal (Q_{surfin}), and outflows in the form of outflow to a river ($Q_{surfout}$) and abstractions from a well field (Q_{well}). Equation (4.1) can be worked out for this example as follows:

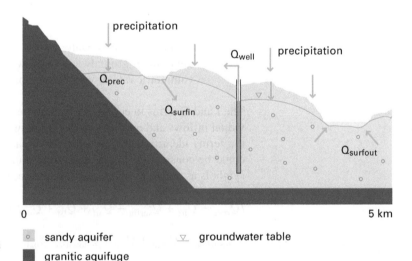

Figure 4.2. Cross section showing a sandy aquifer with groundwater balance terms.

$$[Q_{\text{prec}} + Q_{\text{surfin}}] - [Q_{\text{surfout}} + Q_{\text{well}}] = S_{\text{bal}} \qquad (4.3)$$

Equation (4.3) shows that only five terms have to be considered in the groundwater balance. A low number of groundwater balance terms does not only reduce field activities, but also enhances the reliability of the balance. When more terms have to be considered inaccuracies are being introduced.

The equilibrium groundwater balance

A non-equilibrium groundwater balance such as equation (4.1) has a term that needs further attention: the volumetric rate of water stored or released (S_{bal}). This term relates to equations (3.46) and (3.47). The analogy indicates that for a selected observation period, the water stored in or released from the groundwater balance area relates to changes in hydraulic head. A net increase in hydraulic head between the start and end of the observation period means that water is stored in the area, and in case the head decreases, water is released. In case there is no change in hydraulic head, then water is neither stored nor released. The S_{bal} term in the groundwater balance is then zero and the balance is called the 'equilibrium groundwater balance'. The equilibrium groundwater balance, for instance, for the illustrative example shown in Figure 4.2, can be written as:

$$[Q_{\text{prec}} + Q_{\text{surfin}}] - [Q_{\text{surfout}} + Q_{\text{well}}] = 0 \qquad (4.4)$$

Preferably one tries to set up equilibrium groundwater balances. They can be more easily determined than non-equilibrium groundwater balances. When an observation period is selected for which the balance has to be determined, then one could check whether the hydraulic heads, measured as groundwater levels in observation wells, are the same, at the start and at the end of the period. Then, S_{bal} can be neglected. For example, in many places, the hydraulic heads show a yearly seasonal fluctuation, whereby high heads are observed in the rainy season and low heads are measured in the dry season. In that case, a so-called 'hydrological year' can be defined as 'the period from the end of a dry season to the end of the next dry season'. Provided that the hydraulic heads at the ends of successive dry seasons are similar, then the hydrological year could be considered as a suitable observation period for the determination of equilibrium groundwater balances.

General relationship between terms of the groundwater balance and hydraulic heads

Terms of the groundwater balance affect the hydraulic heads in a groundwater system. Spatial and temporal variations of the inflow and outflow rates of the balance terms are reflected in the hydraulic heads. A general formulation to express the effect of groundwater balance terms on the hydraulic heads in a groundwater system could be as follows:

$$\phi = f\left(Q_{\text{prec}}, Q_{\text{irr}}, Q_{\text{surfin}}, Q_{\text{surfout}}, Q_{\text{well}}, \text{etc.}\right) \qquad (4.5)$$

where:
ϕ = hydraulic head (m)

The inflow and outflow terms of the groundwater balance affect the hydraulic heads in a number of ways. Reference is usually made to 'average, periodic, and trend-like' effects. Long term average recharge and discharge rates affect the average hydraulic heads in a groundwater system. Larger recharge rates will lead to higher values of the average hydraulic heads. Seasonal or short-term fluctuations in recharge rate including recharge from precipitation (Q_{prec}) influence the periodic behaviour of the hydraulic heads. Seasonal variations in abstraction rates (Q_{well}) also affect the periodic behaviour of hydraulic heads. Sudden or gradual increases or decreases in recharge or discharge rates will have a trend-like effect on hydraulic heads. Sudden changes may lead to series of hydraulic heads showing so-called 'step trends' and gradual changes may be reflected as 'linear trends'.

Figure 4.3 presents an illustrative case of the effect of variations in the terms of the groundwater balance on hydraulic heads. The case concerns an area located near the German border in the eastern part of The Netherlands where the groundwater system consists of an aquifer made up of a sequence of fluviatile and periglacial medium to coarse-grained sands and occasional gravels. The graphs in the figure show the relationship between, on one hand, the rate of recharge from precipitation (Q_{prec}) and the abstraction rate at a well field (Q_{well}), and on the other hand the hydraulic heads. The heads were measured as groundwater levels in an observation well nearby the well field. The long-term recharge rate into the system, and the open water levels at local streams are, to a large extent, responsible for the average hydraulic heads in the system. At the observation well, these heads are in the order of 19.25 m above sea level. The yearly seasonal fluctuation in recharge from precipitation results in a 12-month seasonal periodicity of the hydraulic heads. The effect of precipitation on hydraulic heads has a delay of about 1 to 2 months. Although a linear downward trend in the abstraction rate appears to be indicated, a similar trend cannot clearly be noticed on the graph representing the hydraulic heads at the observation well.

4.1.2 *Practical application of groundwater balances*

Why are groundwater balances compiled?

Groundwater balances are compiled for a multiple of reasons. Some of the most important reasons are, that balances:

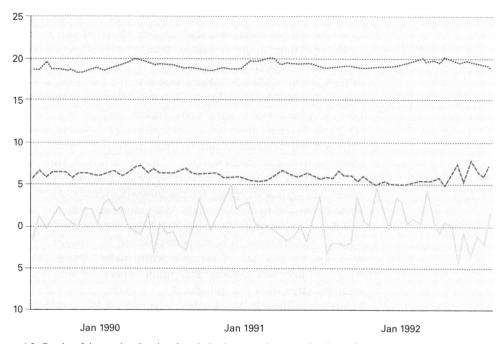

Figure 4.3. Graphs of time series showing the relation between the rates of recharge from precipitation (as positive values for rainfall minus evapotranspiration, indicated by parts of the yellow line; scale indicates mm/day), abstraction rates (red line; scale indicates m³/day times 500), and hydraulic heads (dark blue line; scale in m above mean sea level). The time series are based on half-monthly measurements in the Weerselo area near the German border, The Netherlands (Wu, 1997).

- *Contribute to the understanding of the hydrogeology of an area.* The identification of groundwater inflows and outflows discloses the flow characteristics of a groundwater system.
- *Assist in groundwater modelling.* Inflow and outflow terms of the groundwater balance are input data for groundwater models. Groundwater balance terms are also used for the optimisation of other model parameters during model calibration (see section 3.2.2). On the other hand 'faulty' estimates of the terms of the groundwater balance may be identified through modelling.
- *Assist in water supply studies.* Groundwater balances form the basis for assessments on available groundwater resources. Insight into the terms of the groundwater balance gives an idea on the amounts of groundwater that can be abstracted from groundwater systems for water supply purposes (see section 6.3).
- *Are used to predict effects on the natural environment.* In a natural environment the activities of mankind influence the groundwater balance. Activities may include deforestation and cultivation, redistribution of river water, or installation of abstraction wells and drainage systems. Groundwater balances are set up to predict the effect of 'new or changed' balance terms on 'the other terms'.

Practical example

Consider a practical case of the application of the groundwater balance concept. For a water supply study, the effects of well abstractions on the environment are evaluated. What can one expect when wells are installed to abstract large amounts of groundwater? Knowledge of the local groundwater balance may give the following answer. By introducing abstractions (Q_{well}), one may expect that the other outflows will be reduced. These could include outflow to surface water ($Q_{surfout}$), capillary flow (Q_{cap}), springflow (Q_{spring}), etc.

A second reaction could be that the inflow terms including recharge from precipitation (Q_{prec}), are not affected. However, this is not entirely true in all areas. One could imagine that, to some extent, the inflow increases for the following reason. As a result of abstractions the hydraulic heads (the groundwater levels) decrease and this could increase the typical recharge area of the groundwater system. The increase in recharge area could be brought about by a reduction, for example, of the area, where water logging and surface water outflow takes place. The result would be that the inflow term, recharge from precipitation (Q_{prec}), in the groundwater balance increases. Although the sketched example could be realistic for a large number of groundwater systems, there are no fixed rules with respect to predicting the influence of abstractions on groundwater balances. Each groundwater system and each abstraction has its own set of typical consequences for the groundwater balance that have to be assessed through careful study.

4.2 GROUNDWATER BALANCE ESTIMATION METHODS

4.2.1 *Basic concepts and recharge*

Basic concepts

Various methods are in use to estimate the inflow and outflow terms of the groundwater balance. They may include direct methods or rather indirect methods based on groundwater flow computations or modelling. However, one will have to realise that not all the inflows and outflows can be estimated with the same accuracy. For instance, it is generally considered a difficult task to make reliable estimates for the recharge from precipitation (Q_{prec}), or for the capillary flow from shallow groundwater tables (Q_{cap}). In other areas abstractions from wells (Q_{well}) may be difficult to estimate when no proper records have been kept. Obviously, one attaches less 'weight' to the inflow and outflow terms of the groundwater balance that can be estimated with less certainty. In some cases one does not try to determine them at all. They are just calculated as the 'unknown term' in the groundwater balance.

Consider an illustrative case whereby one focuses on the determination of certain terms of the groundwater balance, while other terms are hardly considered. Figure 4.4 shows a groundwater system consisting of an aquifer of permeable karstic limestones surrounded by impermeable shales and granite. Recharge originates from surplus precipitation (Q_{prec}) and discharge is in the form of outflow into a river ($Q_{surfout}$) and as spring discharge (Q_{spring}), which is used for town supply. In the area, there is no recharge caused by irrigation (Q_{irr}), no abstraction by wells (Q_{well}), and no capillary flow from shallow groundwater tables (Q_{cap}). The following equilibrium groundwater balance for the limestones can be formulated (see also equation 4.4):

$$[Q_{prec}]-[Q_{surfout} + Q_{spring}] = 0 \qquad (4.6)$$

Although new techniques to obtain better estimates may still be developed, it is well known that, for limestone areas, the recharge due to precipitation (Q_{prec}) is difficult to determine. The data that are usually available allows one to make only rough estimates of the recharge rates. The groundwater discharge at springs (Q_{spring}) and outflow rates into the river ($Q_{surfout}$) can be determined with sufficient accuracy on the basis of records for spring and river discharges. The strategy for

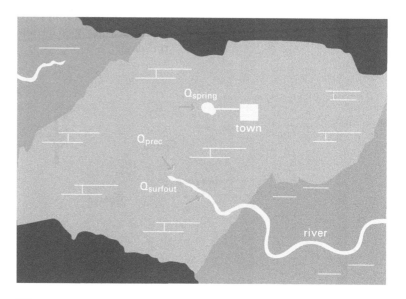

Figure 4.4. Map showing the groundwater balance terms for a limestone aquifer.

■ granite aquifuge

▨ shaley aquifuge

▥ karstic limestone aquifer

estimating the groundwater balance terms would be to focus on these discharges and to pay less attention to the determination of the recharge term.

Recharge from precipitation or irrigation

Three methods will be discussed for the estimation of recharge from precipitation (Q_{prec}). In principle the methods can also be used to estimate the recharge from irrigation losses (Q_{irr}):

1) W*ater balance method.* With this method the recharge rate is estimated on the basis of the water balance for the root zone (see section 1.2.4). The water balance for this zone above a groundwater balance area, as shown in Figure 4.5, can be formulated. In case water storage changes above land surface can be neglected between the start and end of the selected observation period, then the balance can be expressed as:

$$[(P - E_i - E_{pond} - R) + Q_{cap\text{-}root}] - [E_{soil} + T + Q_{perc}] = S_{root} \quad (4.7)$$

where:

P = precipitation rate (m³/day)
E_i = evaporation rate of intercepted water in the vegetational cover (m³/day)
E_{pond} = evaporation from ponded water (m³/day)
E_{soil} = evaporation rate from the soil surface (m³/day)
R = surface runoff (m³/day)
Q_{perc} = rate of percolation at the lower boundary of the root zone (m³/day)

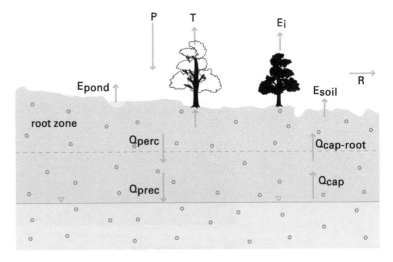

Figure 4.5. Cross section through the unsaturated zone showing the water balance terms.

° unsaturated aquifer   ▽ groundwater table

° saturated aquifer

$Q_{cap\text{-}root}$ = rate of capillary flow at the lower boundary of the root zone (m³/day)

T = transpiration rate of vegetation (m³/day)

S_{root} = volumetric rate of water stored or released in the root zone (m³/day)

Assume that one restricts oneself to recharge areas where there is no capillary flow from shallow groundwater tables (Q_{cap} and $Q_{cap\text{-}root}$ are zero). Lumping also the evapotranspiration terms together as $E = E_i + E_{pond} + E_{soil} + T$, then equation (4.7) can be simplified and re-arranged as follows:

$$Q_{perc} = P - E - R - S_{root} \tag{4.8}$$

Equation (4.8) yields the percolation rate at the lower boundary of the root zone which allows one to make an estimation of the recharge rate from surplus precipitation (Q_{prec}). One can carry out the water balance computations using data on the root zone that are determined on a daily or weekly basis for a sufficiently long monitoring period. In other words a 'running balance' on a daily or weekly basis can be kept. The running balance includes data on precipitation rate, potential and actual evapotranspiration rate, runoff and water content of the root zone, and can be recorded and assessed at hydrometeorological stations. In addition to the 'running balance' method, also models of the unsaturated zone are nowadays prepared to compute recharge from precipitation. These include so-called 'lumped parameter' models (Gehrels, 1999) and numerical models which take into account both the flow and the water balance in the root zone (De Laat, 1980). For the running balance method and the modelling techniques, the determination of the actual evapotranspiration is often a difficult task. This is the principal weakness of these methods to determine percolation and recharge rates.

2) *Chloride method.* This method combines the water balance of the root zone (see above) and the transport of chlorides through this zone. Chlorides dissolved in precipitation and as dry deposition reach the land surface (see section 5.1). These chlorides may infiltrate into the root zone and be transported downward to the groundwater table. Normally, chlorides on their way through the root zone do not precipitate, nor are chlorides added from other sources. In that case the mass transport rates of chlorides in solution can be formulated on the basis of equation (4.8):

$$c_g\, Q_{perc} = [c_p\, P + D_d - c_e\, E - c_r\, R] - c_s\, S_{root} \tag{4.9}$$

where:

c_g = chloride concentration in percolating groundwater (g/m^3)

c_p = chloride concentration in precipitation (g/m^3)

D_d = dry deposition of chloride (g/day)

c_e = chloride concentration in evapotranspiration water (g/m^3)

c_r = chloride concentration in runoff (g/m^3)

$c_s S_{root}$ = mass rate of chloride stored or released in the root zone (g/day)

To simplify matters, it will be assumed that surface runoff does not occur and that an observation period is selected for which the rate of chloride mass stored or released in the water of the root zone, ($c_s S_{root}$), can be neglected. Obviously, in areas where the runoff cannot be neglected, this parameter should be taken into account which makes the formula slightly more complex. Realising that the chloride concentration in evapotranspiration water is usually very small, then equation (4.9) can be adapted and re-arranged as follows:

$$Q_{perc} = \frac{[c_p P + D_d]}{c_g} \tag{4.10}$$

With this formula, the percolation from the root zone can be computed which allows an assessment of the recharge from precipitation (Q_{prec}). One only needs to know the precipitation, the dry deposition of chloride and the chloride concentrations in the precipitation water and in the percolating groundwater. These components can be measured in the field. Compared with the water balance method, this method does not require information on actual evapotranspiration rates. The chloride method may also be suitable for consolidated (hard) rock areas.

Despite these promising signals, the application of the chloride method is not always as straightforward as it appears. In reality, there may be sources of chlorides other than precipitation, affecting the chloride balance. For example, imagine the dissolution of chloride in root zones containing natural salt deposits and the role that 'chloride-rich' fertilisers or irrigation water may play. These components may change the concentration of chlorides and, if unavoidably present, they should be taken into consideration to produce a more relevant equation for the computation of percolation and the assessment of recharge from precipitation.

3) *Groundwater table fluctuation method.* In this method the recharge from precipitation is estimated from an analysis of groundwater table fluctuations. Recharge causes a rise of the groundwater

table. This rise causes an increase in the amount of water stored in the groundwater balance area. However, the water stored is not exactly the same as the recharge, since part of the recharged water is simultaneously discharging from the area. In other words, the recharge is equal to the amount stored plus an amount that is added to discharge. To find a suitable expression for the rate of recharge, (Q_{prec}), the storage equation (3.47) can be considered. Replacing the difference in hydraulic head with a modified rise in groundwater table, as indicated in Figure 4.6, then the following expression can be formulated:

$$Q_{\text{prec}} = F\, S_{\text{y}}\, \frac{\Delta \phi_{\text{mod}}}{\Delta t} \tag{4.11}$$

where:

Q_{prec} = rate of recharge from precipitation (m^3/day)
F = surface area of the groundwater balance area (m^2)
S_{y} = specific yield (dimensionless)
$\Delta \phi_{\text{mod}}$ = modified rise in groundwater table elevation (m)
Δt = observation period (days)

Equation (4.11) can be used to compute the rate of recharge (Q_{prec}) for a specific observation period. In case there is more than one precipitation event within an observation period, then the modified rises can be summed for the calculation of an average recharge rate. The groundwater table rises can be compiled on the basis of measurements in shallow observation wells and the values for the specific yield can be obtained from pumping tests or grain size analyses.

Some remarks, however, can be made regarding this method. First of all, it may be rather difficult to estimate the groundwater table rises correctly. Not only the areal density of the shallow observation wells

Figure 4.6. Graph showing the effect of precipitation on the groundwater table. The modified rises in groundwater table are indicated by: $\Delta \phi_{\text{mod}(1)}$, $\Delta \phi_{\text{mod}(2)}$ and $\Delta \phi_{\text{mod}(3)}$.

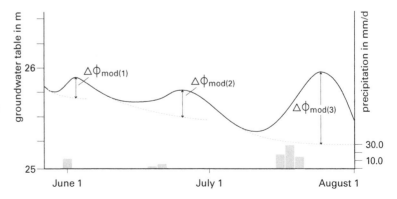

where groundwater tables are measured should be sufficient to obtain representative average values for the rise, but also the measurement frequency should be adequate. Taking the measurement frequency too large could mean that the groundwater table rises caused by the smaller precipitation events are not noticed. Perhaps one should decide to resort to the continuous measurement of groundwater tables. Other disturbing effects may also prevent a sound analysis of the groundwater table rises. For example, abstractions in the area or other human practices could influence the rises and their effect may have to be corrected for. On the other hand, in areas with deep groundwater tables, the dampening effects in the unsaturated zone may hardly cause a response of an individual recharge event on the groundwater table.

Secondly, a representative specific yield for the balance area may be difficult to determine. Usually, values are only available at some points in the area. Averaging these local values to obtain a representative value for the whole area may be quite a risky undertaking. In particular in hard rock areas where there is a large aerial variation in specific yield 'averaging' over only a small number of values may lead to incorrect results (Sibanda, 2009).

Inflow from surface water

How does one determine inflow from surface water (Q_{surfin})? First of all, the type and nature of the surface water will have to be assessed. Is one dealing with a small stream or a wide river, a canal for water supply or for navigation, or is one considering a lake? Their potential to provide substantial inflow to the groundwater system can be quite different whereby the permeability of the bottom material plays an essential role.

Various methods can be followed to estimate groundwater inflow rates from streams, rivers and canals. One of the methods is based on the detection of differences in discharge. Differences will be noticed when one subtracts from each other, the discharges recorded in two or more measurement stations set up along a watercourse. Detected differences for a stream, river or a canal may indicate an inflow of water into the groundwater system.

Nevertheless, one has to be careful with the application of the method. Differences in discharge may also be attributed to water that is diverted at intakes for domestic water supplies or irrigation. In practice this would mean that a complete surface water balance has to be compiled for the stream, river or canal section between the considered measurement stations. The inflow into the groundwater system, being the unknown term, can then be calculated from the balance.

Figure 4.7 shows a fine example of an inflow of surface water into a groundwater system. The picture shows a small influent stream on the island of Sulawesi in Indonesia. The stream 'looses' its water into coarse-grained aquifer material filling the exposed river valley.

One of the methods to estimate the inflow of water from a lake into a groundwater system is based on the compilation of a lake water balance. This inflow is one of the terms of this water balance. Other balance terms include the inflow from direct precipitation on the lake surface, from streams and rivers or from surface runoff, and the outflow in the form of evaporation from the water surface and, also, as streams and rivers. Changes in the volume of water stored in the lake may also be part of the lake water balance. The unknown inflow to the groundwater system can be computed if all the other terms of the lake water balance are known.

Figure 4.7. View of a small stream recharging the groundwater in Sulawesi, Indonesia.

4.2.2 *Discharge and storage*

Discharge to springs

The best method to determine groundwater discharge to springs (Q_{spring}) is by direct measurement in the field. Periodic measurement of discharges with a flow meter or continuous recording of water levels which are converted to discharges, supply all the necessary information. Water levels may be recorded at weirs or flumes set up in an outlet channel near a spring. Provided there is no influence of surface runoff, the combined discharge of a group of springs can also be measured at a site downstream the springs. Especially where large concentrated springs are emerging from a groundwater system, the regular measurement of spring discharges is efficient and useful.

Figure 4.8 shows a spring near Jericho in the Middle East, emerging from the Cenomanian limestones and dolomites west of the Jordan

Figure 4.8. View of a large spring emerging from karstified limestone near Jericho.

Valley area. Measurements taken at the spring with a flow meter on a monthly basis indicate a substantial discharge over 100 l/s.

Discharge through capillary flow

The capillary flow is perhaps the most difficult term to determine in a groundwater balance. Nevertheless, the so-called 'groundwater table depth method' has been used for the estimation of the capillary flow (Q_{cap}). In this method one estimates the rate of capillary flow from the depth to the groundwater table, the grain sizes in the unsaturated zone, and the potential evapotranspiration (see section 1.2.2). Experiments can be carried out to establish the relationship between these parameters. The outcome of these experiments is usually presented in diagrams showing the relationship between the rate of capillary flow or upward flow rate, and the depth to the groundwater table below the root zone for a variety of grain sizes (Doorenbos & Pruitt, 1977). A diagram also represents a particular set of meteorological conditions. Using the diagrams, one can simply read off the capillary flow for measured values of the groundwater table depth and the analysed grain sizes.

Outflow to surface water

How does one assess the outflow of groundwater to surface water including lakes, streams, rivers, and canals ($Q_{surfout}$)? One could apply methods similar to those used for the determination of the inflow from surface water (see above). However, it is useful to point out another method that can be selected to estimate the outflow of groundwater to surface water. The method is based on the separation of hydrographs of streams or rivers into contributions relating to groundwater outflow and to surface runoff (see also section 1.2.3.).

To illustrate the method, the hydrograph in Figure 4.9 can be considered. This hydrograph covers the discharge of a river during two dry periods and one wet period. The peaks in the hydrograph during the wet period represent surface runoff events as a result of precipitation. The 'blank' area under the hydrograph is the surface runoff contribution, while the 'hatched' parts are the groundwater outflows of the current and previous wet periods. The line separating the surface runoff and groundwater outflow for the current wet period is drawn quite arbitrarily, but methods are available to carry out the separation in a more accurate way (Wilson, 1983; Gonzales, 2009). The groundwater outflows shown in the graph can be estimated and related to the outflows into the river, upstream of the site where the discharge measurements are taken. However, when carrying out the analyses one should be aware that part of the groundwater contributions shown on the hydrograph, may also represent discharge from springs. To estimate the $Q_{surfout}$ properly, the discharge from springs (Q_{spring}) should be subtracted.

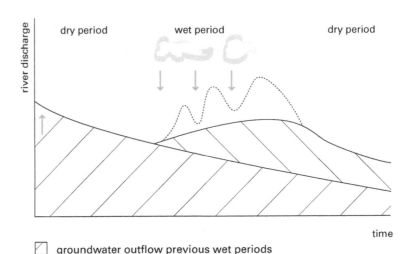

Figure 4.9. River hydrograph
showing the surface runoff and
groundwater contributions.

*Groundwater
abstractions*

How does one assess groundwater abstractions at wells and well fields (Q_{well})? There are various methods in use that include extensive field measurements. A selection of these methods is as follows:

– *Methods based on direct measurements.* Direct and continuous measurement of the abstractions at wells can be carried out using watermeters. Readings taken at the watermeters can be converted into daily or monthly abstractions.

– *Methods based on indirect measurements.* Periodic measurement of the flow of water produced by wells using flow meters, orifice weirs or V-notch weirs can be combined with the registration of pumping hours (e.g. through electricity records). On the basis of the periodic measurements and the registered pumping hours, daily or monthly abstractions can be computed.

The methods outlined above are normally applied at wells where large supplies of groundwater are abstracted for domestic, industrial or irrigation purposes. Legislation enforces the owners of these wells to install watermeters or register pumping hours. Wells used for small farms and household supplies are usually not equipped with watermeters, or are otherwise monitored, and the above methods can not be used. By estimating the number of wells on the basis of field surveys, topographic maps and aerial photographs, and by estimating the yield of selected wells total daily or monthly groundwater abstractions for these wells can also be estimated.

Groundwater storage

Water stored in, or released from a groundwater balance area (S_{bal}) can be determined by 'setting up convenient groundwater storage and release relations'. Mathematical relations that closely resemble equations (3.46) or (3.47) can be established to complete the computations for groundwater storage or release. The appropriate expression for confined and semi-confined groundwater can be formulated as:

$$S_{bal} = GS \frac{\Delta\phi}{\Delta t} \qquad (4.12)$$

where:
S_{bal} = volumetric rate of groundwater stored or released (m^3/day)
G = surface area of the groundwater balance area (m^2)
S = storativity (dimensionless)
$\Delta\phi$ = difference in hydraulic head (head as piezometric groundwater level) between the start and end of the observation period (m).
Δt = length of observation period (days)

For unconfined aquifers one can write the following expression:

$$S_{bal} = GS_y \frac{\Delta\phi}{\Delta t} \qquad (4.13)$$

where:
S_y = specific yield (dimensionless)
$\Delta\phi$ = difference in hydraulic head (head as groundwater table elevation) between the start and end of an observation period (m)

The differences in hydraulic heads can be obtained from measurements in observation wells (see also sections 3.1.1 and 4.2.1). Values for the storativity for confined or semi-confined aquifers and values for the specific yield for unconfined aquifers can be determined from pumping tests. In case the storativities or specific yields vary considerably in the groundwater balance area, one prefers to compute the water stored or released for homogeneous sub-areas first and then one takes a 'weighted' average for the whole area.

4.2.3 *Groundwater balances using flow computations and modelling*

Groundwater balances and flows

It is not always easy to estimate inflow and outflow terms for the groundwater balance using the methods described above. A complex hydrogeology, difficulties in field measurements or simply the lack of data can make it quite difficult to estimate the terms of the balance with sufficient accuracy. In those cases one may assess the groundwater balance by other methods that are based on an indirect estimation of one term or a set of combined terms of the groundwater balance. The best-known methods focus on groundwater flow computations and groundwater modelling.

The estimation of the terms of the groundwater balance by flow computations simply starts with 'getting familiar' with the inflow and outflow terms, and the flow in the groundwater system itself. When the flows are understood, flow computations can be done to quantify the balance terms. In order to carry out the computations, the flow in the area of interest is schematised which usually means that the flow in the aquifers is assumed to be horizontal and in the aquitards vertical flow prevails. Methods including 'simple computations' in aquifers or aquitards, and the flownet approach can then be applied to compute the required terms of the groundwater balance (see section 3.2).

Figure 4.10 shows an illustrative case of the use of groundwater flow computations to estimate inflow terms of the groundwater balance set up for the Bannu Basin in Pakistan (Khan, 1968). In the area, a groundwater system consisting of gravel and sand aquifers with clayey

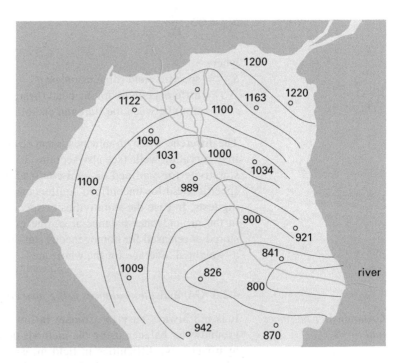

Figure 4.10. Groundwater head contour map for the groundwater system in the Bannu Basin, Pakistan. Note that heads are in feet above mean sea level.

sand and gravel aquifer

sedimentary rock

900 groundwater head contour line

aquitards is recharged at its mountainous boundaries by surface water inflow from small streams (Q_{surfin}), and possibly by lateral subsurface groundwater inflow from adjoining consolidated sedimentary rocks (Q_{lsi}). The combined inflow terms can be estimated by computing the groundwater flow in the system 'downstream' of the inflow zones. The figure shows the groundwater head contour map that has been used to determine the groundwater flow using 'simple computations' and the flownet approach. The combined inflow into the Bannu Basin has been computed at about $3*10^5$ m³/day.

The estimation of inflow and outflow terms using groundwater models leans heavily on the sound description of the conceptual groundwater model (see section 3.2.2). The estimation of the balance terms themselves is based on the results of model calibration. In this modelling phase, selected balance terms can be optimised (see section 4.1.1). In practice, the terms selected for optimisation are usually, but not necessarily, recharge from precipitation and capillary flow. Although, the application of a model seems to be the ideal way to quantify the groundwater balance terms, one has to be extremely careful. The model will not produce satisfactory results when it contains too many unknown- or poorly established geometrical and hydrological data.

Groundwater flow computations and modelling may also be used for the determination of groundwater balance terms for which no methods have been presented in sections 4.2.1 and 4.2.2. These terms are in particular the lateral subsurface groundwater inflow (Q_{lsi}) and outflow (Q_{lso}), and the vertical subsurface inflow (Q_{down}) and outflow (Q_{up}), as defined in section 4.1.1. Methods based on 'simple computations' for the flow in aquifers, and the flownet approach (see section 3.2.2) can be used to compute the lateral subsurface inflow and outflow terms. Methods based on 'simple computations' for the flow through aquitards can be applied to estimate the vertical subsurface flows. Groundwater models with their advanced computational facilities may also assist in establishing lateral and vertical subsurface inflows and outflows for any selected groundwater balance area.

4.3 GROUNDWATER BALANCES AND THEIR ENVIRONMENT

4.3.1 *Groundwater recharge and climate*

Poorly recharged groundwater systems in arid areas

On earth, rock complexes with poor recharge from precipitation are mainly situated in arid areas. Typically, arid areas are located in the northern and southwestern regions of Africa, the Middle East, Central Asia, a large part of the Australian territory, the southwestern part of South America, and in small parts of southwest North America.

Figure 4.11 shows the belts with arid areas on earth. The areas themselves may be sub-divided into hyper-arid, arid and semi-arid areas. For example, in northern Africa, the Sahara desert itself can be classified as a hyper-arid area, whereas the bordering belts running through northern Libya, Algeria, Burkina Faso, Mali, South Sudan, etc, can be considered as arid or semi-arid areas.

Precipitation is the most important criterion used for the classification of arid areas. Hyper-arid areas have very low precipitation rates. Average rates of less than 80 mm/year are not uncommon. In addition, precipitation is very erratic: in some years there is no precipitation at all, whereas in other years intense rainstorms deliver precipitation rates far above the average annual rates. Arid and semi-arid areas have higher precipitation rates. Precipitation rates in semi-arid areas with rainfall in the hot summer season may be as high as 600 to 700 mm/year. Figure 4.12 shows a pictural view of a typical semi-arid area in Turkey. The thunderstorm indicated in the picture may release a substantial part of the winter rain.

The relative scarcity of precipitation is not the only feature of arid areas. In addition, other significant characteristics of arid areas are:
– High levels of solar radiation
– High diurnal and seasonal temperature changes
– High evapotranspiration rates
– Strong winds with frequent dust and sandstorms

Figure 4.11. Map showing the main climatological zones of the world. Arid areas are the steppe and desert zones indicated by pale and light yellowish colours (Times Books, 1994).

Figure 4.12. Pictural view of a semi-arid area near Ankara in Turkey.

– Distinctive geomorphology with poorly developed soils
– Sparse vegetation
– Extreme variability of short duration discharges in streams and 'wadis'
– High infiltration rates in channel alluvium
– Large (seasonal) soil moisture and groundwater storage changes

In most arid areas, recharge from precipitation (Q_{prec}) is limited, and other forms of recharge may be absent at all. In arid areas, groundwater is usually recharged by precipitation in an indirect way. This means that precipitation generally does not recharge the groundwater at the site where it reaches land surface, but is channelled to the alluvium at wadis (see above), to lakes or low-lying areas, or to zones where the rock has cracks, is broken or has other openings. At these places, the collected precipitation will infiltrate. In this way, the water 'escapes' evaporation and recharges the groundwater system Although it is believed that recharge in arid areas mostly takes place in this indirect way, one should not overlook the possibility of direct recharge from precipitation, 'on the spot' itself. Note that attentive field observations will disclose and confirm recharge mechanisms in arid areas (see also section 6.2.2).

Rates of recharge from precipitation in arid areas vary widely. There are meteorological, geological and human factors to be considered. When addressing the meteorological influence, precipitation rate is the determining factor in assessing groundwater recharge. Generally, the very low precipitation rates in hyper-arid areas result in no recharge at all, or in very small recharge rates, whereas the higher precipitation rates in arid and semi-arid areas generate substantially higher rates of recharge. The geological factor is also decisive for recharge in arid areas and is further discussed in section 4.3.2. Finally, the human factor is the most unpredictable factor in recharge assessments. For example, agricultural crop cultivation or the urbanisation of large areas may completely change the natural recharge mechanisms and recharge rates in arid areas.

Perhaps the most interesting aspect concerning recharge rates in arid areas are their long-term changes. In that case one is considering changes against a time scale that is comparable with the geological time scale. On a time scale of several thousands, and even millions of years, the climatological conditions may gradually alter. Surprising changes in climate in various parts of the world have been reported, which brought about long-term changes in recharge rate. The best example is the vast amount of water stored in the Nubian groundwater system in Libya and Egypt. This groundwater originates from recharge that is associated with a much wetter climate that prevailed in the Weichselian pluvial period, over 10,000 years ago. The virtual absence of recharge in areas with vast groundwater reserves has led to the introduction of new terminology. These reserves are also referred to as 'fossil aquifers'.

Recharged groundwater systems in humid areas

Well-recharged rock systems, also referred to as 'well-replenished systems', are usually located in humid areas. Within the context of this textbook, humid areas relate to temperate zones and to tropical areas. Figure 4.13 presents a typical view of a tropical, partly cultivated area in Indonesia. Temperate zones are found in North America and Europe, and the tropical areas are mainly located in South America, Africa, and in Asia.

From a climatological point of view, the temperate and tropical areas differ more from each other than the various arid areas (see above). The temperate regions are characterised by moderate precipitation ranging from 500 to 1000 mm/year. There are usually no explicit rainy seasons. Also, the evapotranspiration rates are moderate, with precipitation rates exceeding evapotranspiration rates in winter, while the reverse is true for the summer period. Tropical areas are characterised by very high precipitation rates that may be well over

Figure 4.13. Typical view of a tropical area with rice cultivation and forest, Sulawesi, Indonesia.

1000 mm/year. Wet and dry seasons can usually be distinguished. As a result of the high temperatures, the evapotranspiration is also high, in particular in the dry season.

Recharge from precipitation (Q_{prec}) can be substantial in temperate and tropical areas. Direct recharge can usually be observed in temperate and tropical areas. Precipitation reaches the land surface and infiltrates 'on the spot'. A determining factor in this process is the relative abundance of soils and weathered zones in temperate and especially in tropical areas, as compared to arid areas. Soils and weathered zones 'soak up' precipitation and transfer the water to the underlying consolidated or unconsolidated rocks. Nevertheless, direct recharge is usually not the only form of recharge in temperate and tropical areas. One should not overlook the possibility of recharge from streams and rivers in temperate and tropical areas (Q_{surfin}). Depending on local hydrological conditions, the river levels may be above the groundwater tables, invoking the inflow of water into the adjoining rock.

Also in temperate and tropical areas, meteorological, geological and human factors influence the rates of recharge from precipitation and

surface water to groundwater. Precipitation surplusses favor groundwater recharge in temperate areas in, normally, the winter season,
whereas in tropical areas the main recharge will be supplied in the
rainy season. Geological and geomorphological factors will be further outlined in section 4.3.2. It is mainly the typical balance between
meteorological, geological and human conditions in an area, which
determines the recharge rates, and the quantities of water that will take
part in the groundwater balance.

4.3.2 *Groundwater balances and geology*

*Groundwater
circulation in
metamorphic and
intrusive rocks*

Arid areas. Metamorphic and intrusive rock complexes are present in a
large number of arid areas. The vast basement metamorphic complexes
in southern Africa, Saoudi-Arabia and Australia are all located in dry
areas. Not surprisingly, rates of recharge from precipitation (Q_{prec}) in
metamorphic and intrusive rock complexes are highly correlated with
precipitation rates.

Figure 4.14, for example, indicates that in the metamorphic gneisses
and intrusive granites of southern Zimbabwe, recharge rates vary from
zero to about 5% of the precipitation rates which range from about 400
to 800 mm/year (Houston, 1988). On the average, precipitation rates
less than 400 mm/year do not result in any recharge. This absence of
recharge for low precipitation rates seems to be a general feature for
metamorphic and intrusive complexes in arid areas. The vast areas

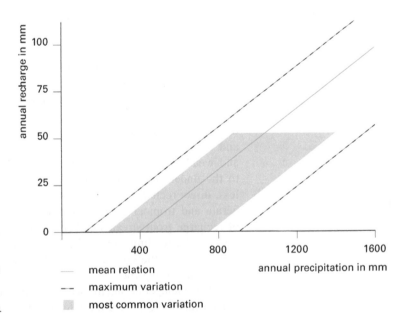

Figure 4.14. Relation between
annual precipitation (rainfall)
and recharge for metamorphic
and granitic rock in Zimbabwe.

— mean relation

-- maximum variation

▨ most common variation

annual precipitation in mm

where these rocks are less permeable combined with low precipitation and high evapotranspiration rates, prevents that recharge takes place.

Groundwater flow and storage in metamorphic and intrusive rocks in arid areas are largely controlled by opened-up fracture systems, by the presence of weathering, the occurrence of 'alluvial sand rivers', and even by the folding pattern (see also section 2.2). There are reports from Botswana that checkerboard types of folding in this country largely control groundwater flow in the basement rock (Gieske, 1992). The discharge of groundwater from metamorphic and intrusive rocks is usually not spectacular. Although small springs may occur, the discharge is usually in the form of capillary flow from shallow groundwater tables (Q_{cap}) or as a modest outflow into a stream system ($Q_{surfout}$). Minor baseflow resulting from this outflow may be measured as surface water discharge.

Temperate and tropical areas. Metamorphic and intrusive rocks also occur in temperate and tropical areas. There are prominent basement complexes in temperate Canada and Scandinavia that largely consist of metamorphic and intrusive rock. Also in tropical areas in Brasil, Central Africa, and southeast Asia these rock types are present. Natural recharge into these rocks mainly originates from precipitation (Q_{prec}). For example, in semi-tropical Karnataka in southwest India, average recharge rates into a weathered and fractured granite complex amount to about 15 to 20% of the average annual precipitation of about 800 mm. For all granites in Karnataka, a much lower recharge rate, in the order of 5% of the precipitation, is common (see also Sinha & Sharma, 1988).

In temperate and tropical areas, the flow and storage of groundwater in metamorphic and igneous rock complexes usually takes place in the weathered parts of the rock, and in particular in the networks of opened-up fractures. Weathered thicknesses of the rock, densities and inter-connections of the fracture systems and the sizes of the individual joints and faults, determine to a large extent groundwater flow rates and storage capacities. Groundwater discharges from metamorphic and intrusive rocks in temperate and tropical areas are in a way similar to comparable rocks in the more arid parts of the world. The discharge usually occurs in the form of small springs (Q_{spring}), capillary flow from shallow groundwater tables (Q_{cap}), and in the form of outflow to streams and rivers ($Q_{surfout}$).

Figure 4.15 shows an example of groundwater discharge in a metamorphic and intrusive rock area. The picture presents an overview of a river branch located in a metamorpic and granitic hard rock complex in Orissa in India. Weathered zones in the hard rock that developed along lines of fracturing and thin, locally deposited coarse-grained

Figure 4.15. Dual groundwater system made up of weathered zones along fractures in the basement complex of metamorphic and granitic rock and pore space in alluvial sediments (shown), Orissa, India.

alluvial material form a shallow groundwater system of limited extent. Recharge from precipitation and irrigation losses, circulate through this system over short distances only and discharge in the form of outflow to rivers ($Q_{surfout}$).

Groundwater transport in volcanic rock areas

Arid areas. Volcanic rocks in arid areas are well known. Some of the vast basalt complexes were introduced in section 2.2. Further examples of volcanic series in arid areas are the Ventersdorp Formation in South Africa, the 'Yemen Volcanics' in arid Yemen, and the Deccan Trap basalts in semi-arid India. Generally, but not everywhere, recharge rates less than 10% of the precipitation are reported for volcanic complexes in arid areas. For example, recharge percentages in the order of 7 to 9% of an annual precipitation of about 600 mm have been assessed for the Deccan Trap basalts (see also Athavale et al, 1983).

The flow of groundwater in volcanic rocks is determined by the presence of opened-up fractures, by connected vesicles, by weathering, and by inclusions of permeable material in between volcanic layers (see section 2.2). In the Deccan Trap area, recharge from precipitation is 'soaked up' by weathered basalt and then flows to depressed areas characterised by fracture systems enabling the transport of water to deeper permeable layers in the volcanic complex. In this way, the basalts are able to circulate appreciable quantities of groundwater.

The discharge of groundwater from volcanic rock complexes in arid areas takes on all sorts of different forms. Large and small springs (Q_{spring}) exist, and there can be capillary flow from shallow groundwater tables (Q_{cap}). Outflow to streams is also common ($Q_{surfout}$). Especially, the occurrence of springs in arid volcanic areas has drawn a lot of attention. Many springs have served small and large communities with water. For example, in semi-arid areas in Yemen, small springs with flows in the order of 0.1 to 5 l/s are present in the 'Yemen Volcanics' which is of Tertiary age. For the far more permeable volcanic basalts of Quaternary age in this country, much larger spring flows, in the order of 5 to 20 l/s, have been reported (Ilaco, 1990). In particular, the presence of permeable zones in alteration with nearly impermeable rocks causes the presence of the small and large springs in volcanic rocks.

Temperate and tropical areas. Have large volcanic rock complexes also been reported in temperate and tropical areas? There are indeed vast volcanic series in the highlands of Mexico, series with a more varying volcanic rock composition in the 'Central Massif' in France, extensive series of basaltic rocks on the Ethiopian and Kenian plateaus, and a variety of volcanic series in the Indonesian archipel. Rates of recharge from precipitation (Q_{prec}) for volcanic rock complexes in temperate and tropical areas are higher than the rates for similar volcanic rocks in arid areas. In addition, recharge in the form of inflow from surface water (Q_{surfin}) into the rock complex may be present in volcanic terrain.

Similar to arid areas, the flow and storage of groundwater in volcanic rocks in temperate and tropical areas are controlled by the presence of connected fracture zones, the occurrence of vesicles, the development of weathering, and the interbedding of sands and gravels. Groundwater discharge is also in the form of springs, capillary flow in shallow water table areas, and outflow to streams and rivers.

For example, on the temperate highlands of Ethiopia, large springs in the order of 50 to 100 l/s have been personally observed, emerging from very porous basalt of Quaternary age. Another example concerns the groundwater outflow in the volcanic catchment of the 'Rio Tepalcetapec' in Mexico. A good correlation was established between the type and the age of the volcanic rocks, and the related groundwater outflow rates (baseflows) at the various branches of the 'Rio' (Ipesa Consultores, 1975). The information available on groundwater in volcanic rock in temperate and tropical areas clearly shows that in the permeable rocks, a significant transport of water takes place from usually poorly defined recharge areas to well-defined discharge areas.

Groundwater balances in consolidated sedimentary rock

Arid areas. Consolidated sediments in arid areas include sandstones and carbonate rocks. Examples of sandstone complexes are the Nubian sandstones in Libya and Egypt (see section 4.3.1), the Tawilah sandstones in Yemen, and the Great Artesian Basin sandstones in Australia. Recharge in sandstones mainly originates from precipitation (Q_{prec}). Recharge rates are usually related to precipitation rates. For example, for the mentioned Tawilah sandstones in the Sana'a Basin in Yemen, a recharge percentage of 3 to 4% of the annual precipitation of about 250 mm has been determined on the basis of a modelling study (NWASA & TNO, 1996).

Groundwater flow and discharge in sandstone complexes in arid areas can be extremely small, but can also be substantial. As a result of local pore space and fine networks of narrow open spaces at bedding plane contacts and fractures, the flow velocities in nearly all sandstone systems are small (see also section 2.2). On the other hand, the storage capacity of sandstones can be large. In case groundwater flow in a sandstone complex located in an arid area is significant, then the discharge is usually in the form of small springs (Q_{spring}), capillary flow from shallow groundwater tables (Q_{cap}) or outflow towards streams ($Q_{surfout}$).

Large carbonate rock complexes are present in arid areas. Perhaps the best-known complexes are located in the Middle East, covering extensive areas in Syria, Jordan, Libanon and Israel. In karstic carbonate limestones, typically all precipitation less evapotranspiration recharges the groundwater system (Q_{prec}). Stream discharges in these areas may hardly be present. For karstic and fractured carbonate limestones and dolomites in arid areas, correlations can be set up between recharge rates and precipitation rates.

Figure 4.16 shows that in the karstified Cenomanian limestone and dolomite rocks, west of the Jordan Valley in the Middle East, relatively modest recharge rates in the order of 0.02 to 0.3 mm/day were initially adopted for a modelling study. The lower rates correlated with the very low annual precipitation in the Jordan Valley, whereas the higher rates related to the far higher annual precipitation of 600 to 700 mm/year in the bordering area with a mountainous character.

The flow of groundwater in karstified and fractured parts of carbonate rock complexes can be rather fast. In limestones with large solution holes, groundwater can travel several kilometres per day. In many places, the discharge of groundwater from karstic limestones takes place in the form of large springs (Q_{spring}). Karstic springs have been

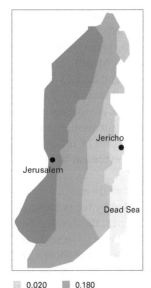

▦ 0.020	▪ 0.180
▦ 0.065	▪ 0.310

Figure 4.16. Computer printout of test recharge rates (mm/day) for a groundwater model study of the groundwater system west of the Jordan Valley, Middle East (Yasin, 1999).

essential for the water supplies of ancient and present day towns. The karstic springs that supply or supplied the towns of Jericho, Amman and Damascus in the Middle East are well known examples.

Temperate and tropical areas. Consolidated sedimentary sandstones and carbonate rocks are also present in the temperate and tropical parts of the world. Examples are the semi-consolidated sandstones of Tertiary age in Germany, the series of chalks in the London Basin, and the limestone or chalky rocks present at many tropical islands. Recharge rates from precipitation (Q_{prec}) in temperate and tropical areas are usually higher than in arid areas. Recharge rates in the order of 15 to 40% are common for parts of the sandstones and limestones in temperate and tropical areas.

The flow and the discharge characteristics of groundwater in sandstones and carbonate rocks in temperate and tropical areas are similar to the flow mechanisms in comparable rocks in arid ares. Networks of narrow open spaces in sandstones and solution holes in karstic limestones and dolomites can be passageways for groundwater flow. The discharge from sandstones is usually in the form of small springs, capillary flow and outflow to streams and rivers, whereas groundwater circulating through limestones and dolomites discharges, in many places, at (large) springs.

Groundwater contained in unconsolidated sediments

Arid areas. Unconsolidated sediments in arid areas are contained in river valleys, and in tectonic and coastal basins. The sediments include gravel, sand, silt and clay. Large unconsolidated rock systems are present in the Mehsana alluvial river valley in west India, in the tectonic basin of the Tihama Plain in Yemen, in the internal basins of northwest Pakistan, and below the coastal plains in the Middle East, Africa, and west Australia. Recharge into unconsolidated systems is usually from precipitation (Q_{prec}), but in many places, one should not underestimate the recharge that can be supplied by surface water (Q_{surfin}). For example, streams that enter an area with unconsolidated rocks may deliver substantial recharge into the sediments.

Rates of recharge from precipitation in unconsolidated sediments in arid areas can be small, but are in most places higher than recharge rates into consolidated rock. Rates have been reported from zero to less than 5% of the average precipitation, but also in the order of 10 to even 25% of precipitation. For example, in the Mehsana river valley, recharge rates in the order of 7.5 to 20% of the precipitation have been determined. For an average precipitation of 700 mm/year, these rates have been estimated using groundwater table rise and groundwater modelling methods (see also section 4.2 and Bradley & Phadtare, 1989).

Figure 4.17 presents an area receiving recharge from precipitation and surface water. The picture shows one of the internal basins

Figure 4.17. View on an internal basin in northwest Pakistan, Domail area, North-West Frontier Province, Pakistan.

of northwest Pakistan with a view across the neighbouring mountain range (see section 4.2.3). The recharge from precipitation into the unconsolidated sediments of the basin itself amounts to less than 5% of the average basin precipitation of about 300 mm/year (WAPDA & TNO, 1983). The recharge supplied by streams from the mountain ranges is substantial and 'outweighs' the recharge from precipitation.

In unconsolidated rocks in arid areas the flow of groundwater through the system takes mainly place through the pore space in gravelly and sandy layers. In most places, the flow of groundwater seeping through silt and clay layers is small, but not insignificant. Groundwater preferably passes through gravelly or sandy 'holes' present in a silt or clay layer. Nevertheless, groundwater flow in unconsolidated sediments advances over a large front.

The main forms of natural discharge of groundwater in unconsolidated rocks are outflow into streams and rivers ($Q_{surfout}$), capillary flow from shallow groundwater tables (Q_{cap}), or flow to springs (Q_{spring}). In many places, springs emerging from unconsolidated sediments are less spectacular than springs that are associated with consolidated rocks. Groundwater which is passing through unconsolidated rock usually emerges at land surface over fairly large 'seepage-like' areas. Well-defined springs are not visible, but water over a wider area is collected and may eventually form the 'head' of a stream.

Temperate and tropical areas. Sediments consisting of unconsolidated rocks are also contained in vast river valleys and basins located in temperate and tropical areas. River valley systems like the Chayo Phraya system in Thailand, a tectonic basin like the 'Central Graben' area in South Germany or the coastal plain system in The Netherlands have been intensively studied. Natural recharge into these valleys and basins originate from precipitation (Q_{prec}) and inflow from surface water (Q_{surfin}). Rates of recharge from precipitation are usually over 20% of the precipitation rates, and may even be as high as over 50% of these rates.

Figure 4.18 shows a diagram with recharge rates from precipitation into the sandy fluviatile and coastal sediments in the northeastern part of The Netherlands. The recharges determined through model calibration amount to 0.2 to 1.2 mm/day, which equate to 10% to 58% of the annual average precipitation of 750 mm/year (Zhang, 1996). The lower percentages can be correlated with urban and forested areas,

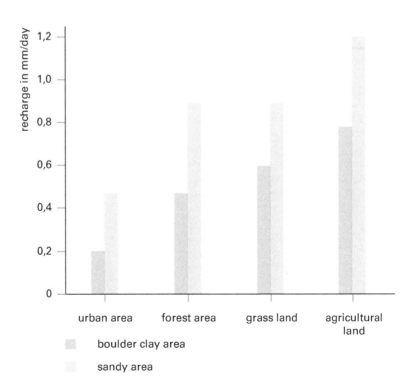

Figure 4.18. Graph showing model calibrated recharges for the sandy fluviatile groundwater system in the northeastern part of The Netherlands.

whereas the higher percentages have been computed for agricultural areas where crops are only grown during a limited period per year. The lower recharge percentages for the glacial boulder clay areas reflect the locally reduced permeabilities in these areas.

High groundwater flow and discharge rates are common for unconsolidated rock systems in the temperate and tropical parts of the world. The flow and discharge mechanisms in these systems are similar to those in arid areas. However, the unconsolidated systems in temperate and tropical areas are usually more prominent, and also have larger storage capacities. Both, small and extensive unconsolidated groundwater occurrences have been thoroughly exploited in many parts of the world.

CHAPTER 5

Groundwater Chemistry

5.1   GROUNDWATER AND CHEMICAL PROCESSES

5.1.1   *Chemical properties of water and rock*

The water molecule

On earth, the water molecule (H_2O) occurs in vast quantities in three states: the solid, liquid and gaseous state. In a groundwater system one finds the water molecule nearly everywhere, as water, in the liquid state. Nevertheless, the water molecule is present, as ice, in a solid form in frozen soils or permafrosts, which cover large regions characterised by a polar climate. A gaseous state of the water molecule may be observed at thermal springs, which often spout enormous amounts of water vapour into the air. In all of its three states the water molecule demonstrates outspoken properties. Water itself can act as one of the most effective solvents. Water also has a high surface tension, a high heat retention capacity, and a high heat requirement for evaporation. Unlike other substances water increases in volume when it freezes into solid ice.

The distinctive properties of water can be related to the structure of the water molecule. This molecule consists of hydrogen and oxygen atoms that have formed a polar bond. A polar bond will be formed when atoms combine into a molecule with an unbalanced electrical arrangement. To understand the formation of such a bond one will have to look at the internal structure of the atoms within the molecule. Atoms of hydrogen and oxygen consist, like all other elements, of a core of positively charged protons, and shells around the core with negatively charged electrons. The atomic number gives the number of protons in the core. For hydrogen and oxygen, Table 5.1 presents the number of protons and the distribution of electrons, in the core and shells.

Atoms combine to form a molecule when they have the option to exchange electrons in such a way that they fill up or empty their shells. The K shell close to the core is full when it contains two electrons and the L shell farther away from the core is full when it holds eight electrons. Thus, hydrogen will form a bond when it either gains an electron or loses its single electron in the K shell. Oxygen will

Table 5.1. Internal structure of the hydrogen and oxygen atoms.

Element	Symbol	Number of protons Core	Number of electrons K shell	L shell
Hydrogen	H	1	1	0
Oxygen	O	8	2	6

combine successfully when it gains two electrons to fill its L shell. One can imagine that two hydrogen atoms are needed to supply the two electrons to the oxygen atom to fill up its L shell to the ideal eight electrons. The combination of two hydrogen atoms with one oxygen atom is stable and explains the chemical formula for the water molecule: H_2O.

Figure 5.1 shows that the eight electrons in the L shell of the oxygen atom in the water molecule are organised into four electron pairs. Two pairs of electrons consist of original oxygen electrons and the other two pairs contain electrons delivered by the hydrogen atoms. The two new electron pairs and the other electrons of the oxygen atom give this atom an overall negative charge. On the other hand, the hydrogen atoms will be positively charged. Thus, the water molecule has a negative side where the oxygen is located and a positive side where the hydrogen

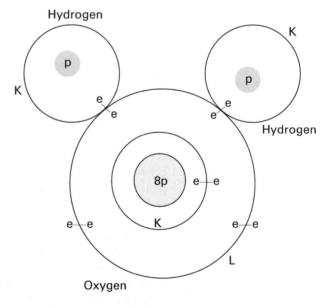

Figure 5.1. Diagram showing the atomic structure of the water molecule.

atoms are placed. The water molecule is therefore electrically unbalanced, forming a polar bond.

The polar bond of the water molecule is responsible for its high di-electric constant and its capacity to form hydrated ions. These properties of the water molecule assist in making water a good solvent (see below). The high di-electric constant of water means that the force of attraction that exists between atoms of a solid substance is easily weakened. Hydration is the property of a water molecule to form a bond with either a positively or negatively charged ion. The negatively charged side of the water molecule can form a bond with a positively charged ion in solution. In the same way the positively charged side of the water molecule may attract negatively charged ions and form a bond in 'aqueous solution'. The high di-electric constant of water and its tendency for hydration play a significant role in groundwater chemistry.

Water as a solvent

Chemical processes play the main role in the release of minerals contained in rocks. Many aspects can be considered, but in this textbook two main dissolution processes will be discussed: the process of hydrolysis (not to be confused with hydration), and the dissolution in acid environments. The process of hydrolysis has been recognised for many centuries and the original meaning of the word is 'the break up by means of water'. This process can actually be described as the reaction between water and a mineral that has the tendency to form a 'weak' acid or a 'weak' base. The chemical reactions that play the main role in hydrolysis are the ionisation of the water molecule, the falling apart of the molecules of the mineral, and the subsequent production of the weak solute. Hydrolysis is the cornerstone in the weathering process of silicate minerals that produce weak silicic acids. The ionisation of water in contact with the mineral can be represented by the following equation:

$$(4)H_2O \Leftrightarrow (4)H^+ + (4)OH^- \tag{5.1}$$

The falling apart of a silicate mineral, for example the mineral forsterite, can be presented as follows:

$$Mg_2SiO_4 \rightarrow 2Mg^{2+} + SiO_4^{4-} \tag{5.2}$$

The formation of the weak silicic acid takes place through the combination of the H^+ and the SiO^{4-}_4 ions. These ions 'like' to combine with each other:

$$4H^+ + SiO_4^{4-} \rightarrow H_4SiO_4 \tag{5.3}$$

Since, the surplus of OH^- ions does not have the tendency to combine, the solution will turn slightly basic. The complete process of hydrolysis may be represented by a single equation that can simply be obtained by adding the three equations above:

$$4H_2O + Mg_2SiO_4 \rightarrow 4OH^- + 2Mg^{2+} + H_4SiO_4 \qquad (5.4)$$

Figure 5.2 shows an area with intense weathering where hydrolysis has played a role. The area is located in volcanic terrain in the highlands of Ethiopia. Local groundwater systems are present in the area. The picture shows that the weathering has resulted in the formation of a loamy soil whereby residual rock fragments set the scene.

The dissolution in acid environments is a process that can be described by the reaction between water and a mineral with the assistance of surplus hydrogen (H^+) ions. A reaction to be considered in this type of dissolution is the formation of acids that deliver the hydrogen ions through ionisation in water. The dissolution takes place in addition to hydrolysis, thereby accelerating the disintegration of minerals. The inevitable example that will be discussed is the dissolution of limestone

Figure 5.2. Mountain area showing cultivated land with large stones. The stones, relocated by the local farmers, are the 'residues' of an intense weathering process of volcanic rock, Debre Tabor area, Ethiopia.

in an acid environment caused by the presence of carbon dioxide. The carbon dioxide is secured through the contact that water and rock have with the atmosphere and through decaying organic material. The formation of the (carbonic) acid can be presented as follows:

$$CO_2 + H_2O \Leftrightarrow H_2CO_3 \tag{5.5}$$

The weak ionisation of this acid to deliver the H^+ ions has the following form:

$$H_2CO_3 \Leftrightarrow H^+ + HCO_3^- \tag{5.6}$$

The tendency for disintegration of the main mineral calcite in limestone, to release calcium and carbonate ions can be represented as follows:

$$CaCO_3 \Leftrightarrow Ca^{2+} + CO_3^{2-} \tag{5.7}$$

The combination of the released H^+ ions and the carbonate ion, to form the weak bicarbonate ion, has the following form:

$$H^+ + CO_3^{2-} \Leftrightarrow HCO_3^- \tag{5.8}$$

The overall reaction can be obtained by adding the various equations. The result clearly shows that carbon dioxide rich water brought into contact with limestone eventually forms free calcium and bicarbonate ions in solution:

$$CO_2 + H_2O + CaCO_3 \Leftrightarrow Ca^{2+} + 2HCO_3^- \tag{5.9}$$

Before addressing the next topic, it should be noted that the reactions above depend on the relative quantity of H^+ ions being present in the solution. In an acid environment with relatively many ions in solution, one definitely has a forward reaction promoting the dissolution of calcium and bicarbonate. However, in an environment with a relative shortage of H^+ ions, the reactions may be reversed, and precipitation of calcite is favored.

Redox reactions

Redox processes that can be represented by redox reactions play a major role in the chemistry of natural groundwater systems. The term redox reaction stands for reduction-oxidation reaction. In these reactions there is a transfer of electrons from one element to another element. The valency of one element is reduced, which is referred to as reduction, whereas the valency of the other element is increased.

The increase in valency is also called oxidation. Reactions describing the reduction part can be formulated. For example, nitrogen with a positive valency may be reduced to nitrogen gas as follows:

$$2N^{5+} + 10e \rightarrow N_2 \text{ (g)} \tag{5.10}$$

The equation shows that by accepting electrons, the valency of nitrogen has reduced. The positive nitrogen atom does not exist freely in groundwater. It may be tied up in a nitrate ion released into ground-water in populated or agricultural areas. The breaking up of the nitrate as part of the formation of the nitrogen gas (denitrification), could be represented by:

$$2NO_3^- + 10e \rightarrow N_2 \text{ (g)} + 6O^{2-} \tag{5.11}$$

Another reaction should be considered to accommodate the released oxygen. The reaction could involve the formation of water as follows:

$$12H^+ + 6O^{2-} \Leftrightarrow 6H_2O \tag{5.12}$$

The last two reactions may be combined to yield the following nitrate reducing reaction:

$$12H^+ + 2NO_3^- + 10e \rightarrow N_2 \text{ (g)} + 6H_2O \tag{5.13}$$

Reactions related to the oxidation part of the redox process can also be established. A fine example is always the oxidation of iron. The iron takes on a higher positive valency to be presented as follows:

$$Fe^{2+} \rightarrow Fe^{3+} + e \tag{5.14}$$

The relationship indicates that the valency of iron is increased by releasing an electron. Also, these types of oxidations occur in groundwater. The oxidation then takes place within the context of a typical chemical process. For example, this may be the precipitation of iron hydroxide:

$$Fe^{2+} + (3)OH^- \rightarrow Fe(OH)_3 \text{ (s)} + e \tag{5.15}$$

The OH^- ion that participates in this process has to be released through another reaction. This could be the ionisation of the water molecule:

$$(3)H_2O \Leftrightarrow (3)H^+ + (3)OH^- \tag{5.16}$$

The complete precipitation process can be described by combining the last two equations above:

$$3H_2O + Fe^{2+} \rightarrow Fe(OH)_3 \text{ (s)} + 3H^+ + e \tag{5.17}$$

Based on the separate reduction and oxidation reactions, the redox reaction itself can be formulated. In nature, a wide selection of reduction and oxidation reactions may be coupled to each other. Assume that the nitrate reduction and iron oxidation as outlined above may also be combined. How does one then formulate the relevant redox reaction? One can simply write down the reaction by combining equation (5.13) with equation (5.17). Equalising the number of electrons involved, one obtains the following overall relationship:

$$24H_2O + 10Fe^{2+} + 2NO_3^- \rightarrow 18H^+ + 10Fe(OH)_3 \text{ (s)} + N_2 \text{ (g)} \quad (5.18)$$

Equation (5.18) shows that the combined nitrate reduction and iron oxidation involves the release of hydrogen (H^+) ions. The produced nitrogen gas dissolves in the water and the iron hydroxide precipitates into the rock material. The reaction occurs when nitrate-rich ground-water mixes with groundwater that is rich in reduced iron.

Another well-known, and closely related, example of a redox reaction is the disintegration of pyrite (FeS_2). In this reaction, the oxidation of both iron and sulphide can be effectuated by the reduction of oxygen (O_2) or nitrate (NO_3^-). The process that triggers off this reaction could be the invasion of oxygen- or nitrate-loaded recharge water into a rock zone containing the mineral pyrite. The process may be catalysed by bacteria that thrive on the energy released in the reaction. Although the reactions will not be elaborated, the final redox reaction of the oxidation of pyrite coupled to the reduction of oxygen will be presented:

$$4FeS_2 \text{ (s)} + 15O_2 + 14H_2O \rightarrow 4Fe(OH)_3 \text{ (s)} + 8SO_4^{2-} + 16H^+ \quad (5.19)$$

For redox assessments, the concept of the 'reduced or oxidised state of groundwater' needs to be introduced. Before this concept can be defined, however, one has to become familiar with the terms 'oxidation potential and redox potential'. The oxidation potential is the voltage that is measured when the oxidation of an element is carried out simultaneously with the reduction of hydrogen. For example, oxidation potentials of zinc, are measured as the voltage between an electrode of zinc, where this metal is oxidised, and a platina electrode, where hydrogen ions are reduced. In groundwater, one does not have one simple reaction like the one for zinc. Instead, a large number of oxidation and reduction reactions are taking place involving an equal number of potentials. In case all the chemical reactions in ground water are known, the potentials of individual reactions could be added to find a combined potential. This combined potential is referred to as the redox potential, represented by the symbol E_h.

In case the redox potential is strongly negative, groundwater is said to be in a reduced state. There is a tendency for elements to take on low valencies. Contrary, if the potential is positive, then water is in an oxidised state. The higher valencies are preferred. Thus, depending on the redox potential, the valencies of the elements are fixed, also determining the form in which elements will manifest themselves in water. The relation between redox potential and the form in which an element is present in groundwater can be visualised with the so-called $E_h - $ pH diagrams. Acidity expressed as pH, being the negative logarithm of the hydrogen (H^+) concentration, has also been selected as a parameter for setting up these diagrams. Redox potential and acidity are, perhaps, the main influencing parameters determining the form in which elements are present in groundwater.

$E_h - $ pH diagrams are compiled for a particular element or a combination of elements. Figure 5.3, presenting the schematic diagram for iron, has been drawn to illustrate the concept of these diagrams. First of all, the $E_h - $ pH area representing the natural groundwater environment, is indicated by the parallelogram in the diagram. The areas where either free oxygen (O_2) or hydrogen gas (H_2) are the stable compounds are also shown. The various stable forms of iron, as dissolved ions or as precipitates with other elements are listed and explained in the diagram. As expected, the lower redox potentials of the reduced state correspond with the lower valencies of iron (Fe^{2+}). The higher potentials of the oxidised state correlate with the higher valencies (Fe^{3+}). The solid lines represent the various oxidation or reduction reactions for iron. For example, moving from the (d) environment to the (b) area across the solid line indicates the oxidation of iron resulting in the precipitation of iron hydroxide. The full reaction is also shown as equation (5.15). Redox diagrams are very powerful tools to assess which chemical compounds are stable for specified E_h and pH conditions in a groundwater system.

Ion exchange

Knowledge of ion exchange is indispensable to understand the chemical composition of natural groundwater. Ion exchange can be defined as 'the replacement of ions at all sorts of surfaces with other ions of similar chemical characteristics.' First, consider the surfaces where ion exchange takes place. In principle, the exchange may take place at all sorts of solid surfaces in the subsurface, but the processes are best known for surfaces of clays, oxides and hydroxides of iron, and organic matter.

Clays, for example, have structures that consist of silicon (Si^{4+}) and oxygen (O^{2-}) ions, in combination with positive ions like aluminum (Al^{3+}), magnesium (Mg^{2+}) and iron (Fe^{2+}). Hydroxyl ions (OH^-) may also be part of the structure. As an illustration, Figure 5.4 shows the basic silica tetrahedron and metal octahedron that are combined in the structure of the clay mineral kaolinite. Since in clays, aluminum ions (Al^{3+}) are basically substitutes for the silicon ions (Si^{4+}), and

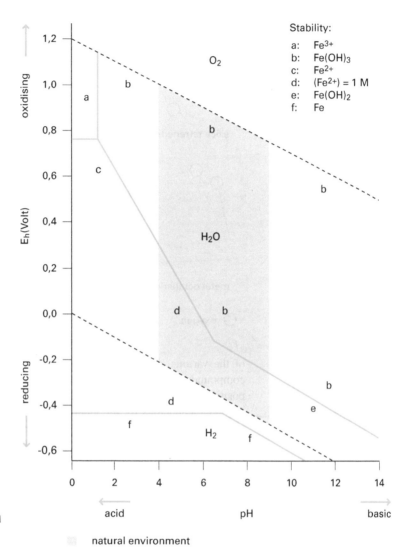

Figure 5.3. The schematic redox diagram for iron. The various forms of iron that are stable in the typical sections of the diagram are indicated (Krauskopf, 1967; reproduced with permission of The McGraw-Hill Companies).

magnesium ions (Mg^{2+}) and iron ions (Fe^{2+}) are replacements for the aluminum ions (Al^{3+}), a deficit of positive valencies exists and positive ions are attracted to the clay mineral. Positive ions that may be linked to the clay surface, are the sodium ion (Na^+), the potassium ion (K^+), and the calcium ion (Ca^{2+}). These positive ions can be exchanged with each other and that is precisely what is meant by ion exchange. Since ion exchange usually concerns the replacement of positive ions, the process is also referred to as 'cation exchange'.

Clays, iron oxides and hydroxides, and organic matter have a different capacity for exchanging cations. The different internal structures

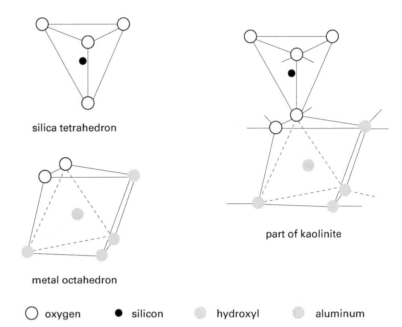

silica tetrahedron

part of kaolinite

metal octahedron

Figure 5.4. Diagram showing the internal structure of kaolinite.

○ oxygen   ● silicon   ◍ hydroxyl   ◍ aluminum

of the various clay minerals, iron oxides and hydroxides and organic compounds are responsible for the different exchange capacities for positive ions. Table 5.2 gives an impression of the different cation exchange capacities for the mentioned solids. The exchange capacity is expressed in milli-equivalents of cation per 100 gram of solid material. For example, the table clearly shows that the exchange at the surface of the clay mineral kaolinite is not large, whereas replacements on surfaces of organic matter can be quite considerable. Other clay minerals like montmorillonite and vermiculite, however, have exchange capacities in the order of 80 to 200 milli-equivalent of cation per 100 gram solid material. These capacities are comparable with the capacities for organic compounds.

Table 5.2. Common cation exchange capacities (Appelo & Postma, 1996).

Material	Exchange capacity (meq/100 gram)
Clay minerals	
Kaolinite	3 - 15
Montmorillonite	80 - 120
Vermiculite	100 - 200
Glauconite	5 - 40
Illite	20 - 50
Iron oxides and hydroxides	up to 100
Organic matter (at pH = 8)	150 - 400

The concept of ion exchange at solid surfaces may have been explained, but how does one formulate the actual exchange reactions? In these reactions the exchange of positive cations with each other is described. For example, groundwater flowing through fractured schists may contain high concentrations of potassium ions (K^+). In case this water subsequently flows through layers with clays consisting of kaolinite rich in sodium (Na^+), the following reaction may take place:

$$K^+ + Na - Kaolinite \rightarrow K - Kaolinite + Na^+ \qquad (5.20)$$

The reaction shows that the sodium at the surface of the kaolinite is replaced by potassium. The reaction may also be reversed which means that the reaction proceeds from 'right to left'. This would happen, for example, where groundwater rich in sodium ions comes into contact with kaolinite clays to which potassium ions are attached.

The cation exchange between potassium and sodium concerns the reaction between two ions with equivalent valency. This need not always be the case. Consider the case whereby groundwater rich in sodium ions (Na^+) invades organic peat layers with calcium (Ca^{2+}). In this case, the sodium ions with a valency of ($+1$) replace the calcium with a valency ($+2$). The reaction showing the cation exchange for the peat layer is as follows:

$$2Na^+ + Ca - Peat_2 \rightarrow 2Na - Peat + Ca^{2+} \qquad (5.21)$$

Relation (5.21) shows the replacement of the calcium in the peat by the sodium. The reaction also indicates that the concentration of sodium ions (Na^+) in the groundwater has decreased, whereas the concentration of calcium ions (Ca^{2+}) has increased. Groundwater leaving the peat layer may have changed considerably in hydrochemical composition.

Chemistry of rocks   Rocks may consist of a single mineral or they may contain more than one mineral. Each mineral has its own chemical formula, which is usually far more complex than the formula for water. Table 5.3 shows the chemical formulae, crystal classes and common colouring of the minerals quartz, orthoclase feldspar, calcite, gypsum, dolomite, and hematite. The table presents only a small selection of the minerals that may be present. In reality hundreds of different minerals do occur in nature, some of which are present in abundant quantities whereas others are rather rare. It is interesting to note in the table, that the selected minerals consist of 'oxides' like in quartz and hematite, 'silicates' as in the feldspars, and 'carbonates' as in calcite.

Table 5.3. Selected minerals in the earth's crust.

Mineral	Chemical formula	Colour	Crystal class
Quartz	SiO_2	Various	Amorphous/hexagonal
Feldspar	$KAlSi_3O_8$ [1]	White/pink	Amorphous/monocline
Gypsum	$CaSO_4.2H_2O$	White	Monocline/fibrous
Calcite	$CaCO_3$	White [2]	Hexagonal
Dolomite	$CaCO_3.MgCO_3$	White	Amorphous/hexagonal
Hematite	Fe_2O_3	Black/red	Amorphous/rhombohedral

[1] Orthoclase feldspar.
[2] Also colourless/other colours.

In section 2.1, rocks are categorised either as consolidated or unconsolidated. Table 5.4 shows a selection of rock types in both categories and the main minerals that one may expect. The chemical formulae of some of these minerals can be found in Table 5.3. For example, it is shown that granite contains quartz and orthoclase feldspars, and clay may consist of clay minerals including illite, montmorillonite and kaolinite. The chemical formulae of feldspars and clay minerals show that these minerals belong to the silicates. The chemical formulae of the minerals making up the rock of a groundwater system are indicators of the type of chemical components that may be present in groundwater (see also below).

5.1.2   *Development of groundwater chemistry*

Basic concepts

The groundwater chemistry in a groundwater system relates to the groundwater balance of the system, the chemical composition of the

Table 5.4. Examples of rock types and mineral content.

Consolidated rocks		Unconsolidated rocks	
Examples:	Main minerals:	Examples:	Main minerals:
Basalt	Plagioclase feldspar, Augite, Olivine	Gravel	Quartz
Granite	Quartz, Orthoclase feldspar, Mica	Sand	Quartz, Feldspars, Mica
Schist	Mica, Chlorite, Hornblende	Loam	Quartz, Clay minerals
Sandstone	Quartz, Feldspars, Mica	Clay	Kaolinite, Illite, Montmorillonite (clay minerals)
Limestone	Calcite		

precipitation and the rocks, the geological history of the area, and the influence of mankind. Groundwater balances in groundwater systems, describing recharge and discharge terms, correlate with flow systems (see also section 3.3). In a flow system recharge from precipitation flows through the unsaturated zone and saturated rock to groundwater discharge areas.

The origin of the precipitation, the physical and chemical conditions of the unsaturated zone and the chemistry of the underlying saturated rock determine to a large extent the chemical composition of the water in a groundwater- and flow system.

The geological history of an area can be very complex and long-term changes in the 'rock-forming' environment may strongly influence the development of groundwater chemistry. For example, rocks that were formed in a marine environment exhibit a typical groundwater chemistry that differs from the chemistry in similar rocks that would have developed under continental conditions. Higher concentrations of dissolved solids are usually present in marine rock. The geological environment in which rocks are formed puts its mark on the chemical evolution of a groundwater system.

Activities of mankind may have disastrous effects on the chemistry of groundwater. These activities include the release of contaminants including fertilisers and manure, pesticides and the disposal of industrial and domestic waste. Inorganic and organic soluble components of these contaminants change the natural chemical composition of the groundwater. To make matters worse, low groundwater velocities in many systems almost 'fix' the contaminants in a groundwater system.

Chemical composition of rainwater

The influence of precipitation and recharge on groundwater chemistry can be further elaborated. To describe the effect of recharge water on groundwater chemistry, the chemical composition of precipitation needs to be assessed. Precipitation contains natural and industrial components. Natural components usually originate from water in oceans and seas and are transported through the atmosphere to continents, as part of the processes playing a role in the hydrological cycle (see section 1.2). Although at much lower concentrations, the chemical composition of precipitation is similar to the composition of seawater.

Elements including sodium (Na) and chlorine (Cl) are usually the dominant natural components in precipitation. Usually these elements are present in dissolved form, but they may also occur as solids. Especially the sodium and chlorine in the solid phase rapidly drop out of the precipitation as so-called 'dry deposition'. The amount of sodium and chlorine in precipitation decreases with the distance from the coast. In The Netherlands, for example, maximum chloride (Cl^-) concentrations at the coast can be up to 30 mg/l, whereas the concentrations have dropped to 3 to 4 mg/l at a distance of about 100 km inland (Ridder, 1978).

The industrial components in precipitation relate to the output of heavy and chemical industries, and mining operations. Especially, the output of sulphur (S) and nitrogen (N) can be substantial. In industrial areas in western Europe, sulphate (SO_4^{2-}) concentrations over 10 mg/l and nitrate (NO_3^-) concentrations over 5 mg/l have been measured in precipitation. These substances turn the precipitation acid and influence the dissolution processes in the groundwater system. On the other hand, when calcite ($CaCO_3$) is released into the atmosphere through industrial operations (cement industry) or excavation activities, the hardness of the precipitation increases and basic conditions will prevail. The industrial components in precipitation are not restricted to small areas whereby major parts of a country remain unaffected. The airstreams in the atmosphere can easily transport industrial outputs over large distances.

Typical processes in the unsaturated zone

Precipitation infiltrates into the soil, passes through the root zone, then through an intermediate zone and finally recharges the groundwater system (see section 1.2). The chemical composition of the recharge water on its way down through the unsaturated zone is influenced by a variety of processes. A selection of the processes that successively play a role in the unsaturated zone is as follows:

– *Increase in concentration by evapotranspiration.* Through direct evaporation and transpiration by plants, the concentration of typical ions dissolved in recharge water will change. An increase in concentration proportional to the share of the recharge water being lost by evapotranspiration can be considered. For example, in case 75% of the recharge water leaves the unsaturated zone by evapotranspiration, and the other 25% is added to the groundwater system, the concentration will increase by a factor 4. Strictly speaking, this 'rule of thumb' may only be applied in case evapotranspiration does not remove ions dissolved in recharge water. Although this assumption is not fully justified (see below), the rule is often used for practical applications.

– *Decay of organic matter.* Plant remains and other forms of decaying organic matter influence the composition of the recharge water. The process can be described by a redox reaction of the following form:

$$CH_2O + O_2 \rightarrow H_2O + CO_2 \tag{5.22}$$

In the reaction, the CH_2O represents the organic matter that is oxidised by free oxygen (O_2). This oxygen is usually, but not always, present in the unsaturated zone. During the process carbon dioxide

(CO_2) is released. The end result of this process is an increase in carbon dioxide concentration in recharge water.

– *Uptake of chemicals by plant roots.* Plant roots may suck up water from the unsaturated zone originating from precipitation. This water is subsequently released by transpiration into the atmosphere. Plants are selective in the chemicals that they take up by their roots. In The Netherlands, for example, maize is planted in agricultural areas since it takes care of the removal of nitrogen compounds from water in the unsaturated zone. Nitrogen, phosphorous and potassium are often reported as being removed by a large variety of plants. The uptake by plant roots changes the concentrations of the chemical ions in the recharge water.

– *Dissolution of minerals.* Rock material in the unsaturated zone can be dissolved by recharge water. In many places, the water reacts with rock material that may have been affected by weathering. The principal reactions that play a role are hydrolysis and dissolution in an acid environment. Silicate minerals, ore minerals, carbonates, etc may fall prone to dissolution in an acid environment created by excess carbon dioxide in solution (see above). When not influenced by other processes, the dissolution increases the rock-related chemical ions in the recharge water.

Processes in the groundwater system

Recharge water from the unsaturated zone entering into and flowing through a flow system in the (saturated) groundwater system is subjected to several processes affecting groundwater chemistry. With the exception of simple mixing, four processes will be discussed. Although the basic theory behind most of these processes is outlined in section 5.1, the discussion below will focus on their application in recharged groundwater systems:

– *Dissolution of minerals along a flowpath.* When recharge water enters a groundwater system, it follows a so-called 'flowpath' through a flow system towards a discharge area. Along a flowpath, the concentration of the ions dissolved in water usually increases (see section 3.3), while the chemical composition of the water also changes. Reactions similar to those in the unsaturated zone take place and include the dissolution of silicate minerals in all types of rocks, calcite in carbonate rocks, sulphate minerals in evaporites, etc. Typical for certain sandy groundwater systems, the chemical composition of the water along a flowpath may subsequently change by dissolution, mixing, and partial precipitation from typical bicarbonate-rich water into sulphate-rich groundwater. Finally, chloride-rich groundwater may be formed. In a schematic way the reactions can be presented as follows:

$$HCO_3^- \rightarrow HCO_3^- + SO_4^{2-} \rightarrow SO_4^{2-} + Cl^- \rightarrow Cl^- \qquad (5.23)$$

– *Redox processes along a flowpath.* Research has revealed that along a
flowpath, the redox potential generally decreases. The redox potential usually relates to the amount of oxygen that is brought into
the groundwater system by means of recharge water. Dissolved in
recharge water, the oxygen may hardly persist, or travel over large
distances into the groundwater system. Three cases will be briefly
reviewed.

In consolidated rocks with large open spaces at joints and faults
or at solution holes one may expect high redox levels caused by a
surplus of oxygen in the recharge water. Surplus oxygen in recharge
water can be anticipated where organic material is lacking in the
unsaturated zone. The redox potentials and oxygen concentrations
remain high over a large distance along a flowpath in the saturated
system. The reason is that the large openings in the rock result in
high groundwater velocities and this limits the time needed for a full
completion of redox reactions.

In unconsolidated rocks high redox potentials and high oxygen
concentrations are also present, where the amount of organic material in the unsaturated zone is limited. When oxygen-rich recharge
water enters the groundwater system, however, a decrease in oxygen
concentration is common in the first tens of meters of a flowpath and
reducing conditions with low redox potentials and oxygen concentrations are established. The change is caused by the low groundwater
velocities in these rock types giving ample time for redox reactions
to take place.

When recharge water has a low oxygen concentration, then the
whole groundwater system may be in a reduced state with low redox
potentials and insignificant oxygen concentrations. Recharge water
devoid of oxygen may be caused by abundant organic matter in the
unsaturated zone, in combination with a fine grain size of the rock
material.

Table 5.5 shows typical redox reactions that may occur in a
ground-water system. The table indicates reactions that could be
invoked in the upper part of an oxygen rich, recharged groundwater
system consisting of unconsolidated sandy material. For example,
sulphides could be oxidised to sulphates and dissolved iron could be
oxidised into solid iron compounds. Further down a flowpath typical reduction reactions could occur, whereby sulphates and iron are
reduced. Even further along the flowpath, methane may be formed.
Bacteria may have to be present to accelerate these processes, or start
them altogether.

– *Ion exchange in the groundwater system.* Groundwater traveling
along a flowpath in a flow system may permeate through rock layers
consisting of clays, iron oxides and organic material. These materials have a high ion exchange capacity. The presence of these layers

Table 5.5. Oxidation processes consuming oxygen.

Process	Reaction	
Sulphide oxidation	$O_2 + {}^1\!/_2 HS^- \rightarrow {}^1\!/_2 SO_4^{2-} + {}^1\!/_2 H^+$	(1)
Iron oxidation	${}^1\!/_4 O_2 + Fe^{2+} + H^+ \rightarrow Fe^{3+} + {}^1\!/_2 H_2O$	(2)
Nitrification	$O_2 + {}^1\!/_2 NH_4^+ \rightarrow {}^1\!/_2 NO_3^- + H^+ + {}^1\!/_2 H_2O$	(3)
Manganese oxidation	$O_2 + 2Mn^{2+} + 2H_2O \rightarrow 2MnO_2\,(s) + 4H^+$	(4)
Iron sulphide oxidation (*)	${}^{15}\!/_4 O_2 + FeS_2\,(s) + {}^7\!/_2 H_2O \rightarrow Fe(OH)_3\,(s) + 2SO_4^{2-} + 4H^+$	(5)

(s) = solid phase.
(*) = expressed as a combined reaction.

may substantially change the chemistry of the groundwater passing through the groundwater system. In particular, the ion concentrations of sodium (Na^+), potassium (K^+), calcium (Ca^{2+}) and magnesium (Mg^{2+}) may be affected when groundwater containing these ions filters through layers with a high ion exchange capacity.

– *Flushing of saline groundwater.* This is a typical process that relates to the geological history of a groundwater system. Consider, for example, the natural development of a fresh groundwater lens in marine unconsolidated rocks accepting recharge water. During the marine deposition of the rocks, groundwater with a saline composition is trapped in the sediment material. After geological uplift, the rocks form a part of a continent and fresh recharge water mixes with and flushes out the saline water. The flushing of saline water transforms the groundwater system into a less saline system, and brackish or fresh water environments are established.

Processes at groundwater discharge points

At locations where groundwater discharges, one may observe the precipitation of dissolved minerals. Groundwater discharge can be observed at springs, at areas with capillary flow, and at stream and riverbeds showing an outflow of groundwater (see section 1.2). Precipitation of minerals in these places is the reverse of the tendency for the dissolution of minerals. An interesting example in this context is the formation of the calcite-rock referred to as 'travertin'. The rock is formed where groundwater saturated with calcium ions (Ca^{2+}) and bicarbonate ions (HCO_3^-) discharges at land surface. The contact of the water with the atmosphere results in a loss of carbon dioxide, which reduces the amount of hydrogen (H^+) in solution (compare with equations (5.5) and (5.6)). The created shortage of hydrogen ions causes the precipitation of calcite in the form of travertin.

Precipitation of minerals in groundwater discharge areas may also occur as a result of an increase in redox levels triggered off by the introduction of oxygen. Figure 5.5 presents an interesting example of iron precipitation induced by an increase in oxygen at a groundwater discharge

Figure 5.5. Top view of brownish iron hydroxide precipitate at a small stream emerging from a culvert, Uden, The Netherlands.

point of an extensive sandy groundwater system in the south of The Netherlands. The picture shows iron hydroxide ($Fe(OH)_3$), precipitated from groundwater discharging through the bed of a small stream. The basic reaction is formulated in Table 5.5 (2). The iron hydroxide can be noticed at the bed, but also at the shown culvert where the water is collected. As a result of aeration at the culvert, additional iron hydroxide is being precipitated which produces the brown iron staining.

Precipitation as a result of high evapotranspiration rates is another process that is often visible in discharge zones in arid areas with high groundwater tables or water logging. Groundwater saturated with chlorides, carbonates, sulphates, etc may become super-saturated by evapotranspiration and minerals like 'rocksalt', calcite and gypsum will precipitate. A special case form areas with excessive irrigation. High groundwater tables, induced by the supply of irrigation water, may give rise to excessive evapotranspiration. In case the groundwater has high concentrations of dissolved solids, precipitation will occur. Crusts of saline precipitates are formed near the contact between the unsaturated zone and the saturated groundwater system, eventually making an area unsuitable for agricultural production.

5.1.3   *Assessment of groundwater chemistry*

Field data collection

The collection of water samples at springs, rivers and wells is an essential activity to unravel the chemical reactions and processes in a particular flow system within a groundwater system. The collection of samples at springs provides an opportunity to assess the water chemistry at groundwater discharge points. Also, the collection of samples at streams and rivers gives insight into the chemistry of discharging groundwater as long as base flows are being sampled and the surface water source has not been affected by pollution. The collection of samples at wells gives the groundwater chemistry at a certain point along a particular flow line within a groundwater flow system. Along different flow lines different groundwater chemistries develop. The depth of the well screen then determines which flow line is being sampled yielding the chance to analyze the related groundwater chemistry.

The taking of water samples at springs, streams and rivers is rather easy and instrumentation or sample bottles can directly be submersed in the water source. At wells the securing of proper water samples is more complex. Taking water samples at a pumping or production well is effective. Instrumentation can be inserted into the pumped water and sample bottles can be filled as long as pumping time has been sufficiently long to abstract proper groundwater from the system. However, when the well is screened over a considerable depth range, a sample is obtained which may contain a mix of groundwater with different chemical signatures. The groundwater chemistry at a particular point along a flow line will then not be properly sampled. In case one indeed requires samples with a unique chemical signature then it is better to lower a water sample collector or use a small submersible pump in an exploration or observation well which is screened at the particular depth of interest in the groundwater system. A multi-piezometer well is an example of an observation well where water samples at different depths can be collected (see section 6.3.3).

The outlet point of the spring portrayed in Figure 5.6 is the scene of an extensive data collection program. The spring emerges from a fractured limestone aquifer which is contained in a folded sequence of sedimentary rock at the outliers of the Alps. In addition to the recording of spring flow, instrumentation has been installed which records the electrical conductivity (EC) of the spring water. Moreover, water samples are frequently collected at the outlet. Not surprisingly, calcium and bicarbonate were analyzed to be the dominant ions dissolved in the spring water.

The sampled water can be analyzed directly in the field or may be taken to a laboratory. Parameters which are easy to measure with instrumentation in the field and may alter during transport should be measured in the field. These parameters generally include the EC, pH,

Figure 5.6 Spring outlet point showing groundwater emerging from fractured Lias limestones, St Benoit, Digne, France. At the bottom right an EC and pH probe together with accessories are shown.

temperature, and dissolved gases like free oxygen. Some specific chemical constituents like bicarbonate should also be determined in the field. Special kits are also available to analyze on site a wide range of parameters. The samples intended to be analyzed in the laboratory should be stored in containers which may be simple plastic bottles, for most of the analyses, up to sophisticated copper tubes for some of the isotope analyses. To minimize the risk of changing the concentration of chemical constituents like calcium (Ca^{2+}), magnesium (Mg^{2+}), sodium (Na^{+}), potassium (K^{+}), and metals (e.g. Fe^{2+}) during storage and transport, duplicate samples should be taken which are filtrated and acidified.

Units and accuracy checks

The concentrations of chemical constituents analyzed in the field, but especially in the lab, are commonly expressed in milligram per liter sample (mg/l) or parts per million by weight of sample (ppm). For chemical computations based on reaction equations the concentrations are usually recalculated into millimoles per liter sample (mmol/l).

The number of millimoles per liter is obtained by dividing the concentration in milligram per liter sample by the gram formula weight of the dissolved ion or molecule in question. To check the accuracy of the analyses and for presentation purposes the millimoles are generally converted into milliequivalents per liter sample. Multiplication of the number of millimoles per liter sample by the charge of an ion delivers the number of milliequivalents per liter sample (meq/l).

Finally, some text should be devoted to checking the accuracy of the analyses. In the first place it may be useful to analyze a selection of the samples in different laboratories to assess the quality of this particular institution. Laboratories that are frequently subjected to fulfill national or international analytical standards should be preferred. Nevertheless, checks following the Electro Neutrality condition and EC comparison should be done to ascertain the accuracy of the analyses (Appelo and Postma, 2009). For the Electro Neutrality condition the ratio of the addition of the sum of major cations and anions, and the subtraction of the sum of these major cations and ions (multiplied by hundred to obtain percentages) is computed. Samples with percentages of more than 5% should be re-examined. For the EC comparison this parameter is computed by formulae or graphs using the analyzed concentrations and compared with field or lab measured values using an EC measuring instrument.

Presenting results   Results of chemical analyses of water samples collected at particular locations in the groundwater system may be presented in tables, and on maps and sections. A popular way to show the results on maps and sections is in the form of so called Stiff diagrams. A Stiff diagram shows, along appropriate scales, the major cations on the left and the major anions on the right. In order to be able to spot the dominant ions at a glance, the units used are expressed in milliequivalents per liter sample (meq/l). Figure 5.7 shows two examples of Stiff diagrams. The left diagram

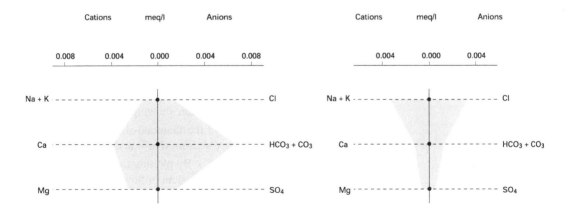

Figure 5.7. Stiff diagrams showing the major cations and anions expressed in meq/l.

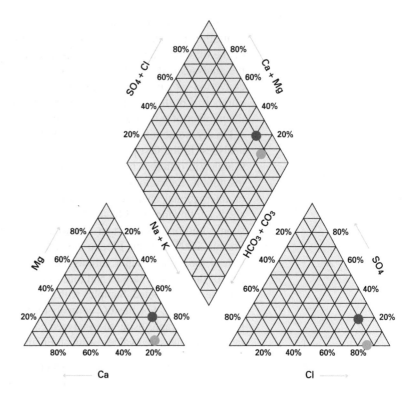

Figure 5.8. Piper diagram showing the analyses of two samples. The distribution of major cations and anions expressed in percentages are based on the use of miliequivalents per liter sample.

indicates that the most important cations analyzed in the sample are calcium (Ca^{2+}) whereas the dominant anions are the bicarbonates (HCO_3^-) and carbonates (CO_3^{2-}). The right diagram shows that the most essential cations and anions are respectively sodium (Na^+) and chloride (Cl^-).

Stiff diagrams show the distribution of cations and anions of a single water sample. A way to show the results of a multiple of samples in one picture is to use Piper diagrams. A Piper diagram consists of two triangles and one diamond-shaped area. Symbols in the triangles show the distributions of major cations and anions analyzed for a selected number of samples. The contributions of the typical ions are expressed as percentages. The diamond-shaped area contains symbols that are obtained by combining the results of the triangles. Using circles as symbols, Figure 5.8 presents a Piper diagram showing the analyses of two selected samples. The triangles and the diamond-shaped area indicate that the most important cations are sodium (Na^+) and potassium (K^+). The dominant anion is chloride (Cl^-). By plotting the sample analyses collected for (a part of) a groundwater system or flow system in a single Piper diagram, the chemical processes playing a role in this system can be interpreted (Appelo and Postma, 2009).

Table 5.6. Classification of groundwater (Fetter, 1994).

Category	Total dissolved solids (mg/l)
Fresh water	0 – 1000
Brackish water	1000 – 10,000
Saline water	10,000 – 100,000
Brines	Over 100,000

The use of 'water types'

Stiff and Piper diagrams can be used to obtain a first impression of the groundwater chemistry in a groundwater system, but the use of so-called water types brings these assessments one step further (Stuyfzand, 1999). The water type is a label indicating, most importantly, the total dissolved solids (TDS) concentration and the main types of dissolved ions in the groundwa ter. For the TDS concentration a classification can be made into fresh, brackish, saline and super-saline water. Table 5.6 shows this classification. The main types of dissolved ions usually include the main positive and negative ions. The notation S-NaCl, for example, means that the groundwater is saline and that sodium (Na^+) and chloride (Cl^-) are the principal positive and negative ions. The introduction of water types has assisted tremendously in analysing the chemical behaviour of groundwater systems.

Figure 5.9 shows an illustrative example of the flushing of saline groundwater and the use of water types. Similar to Figure 2.16, the cross section in Figure 5.9 shows the groundwater system underlying the coastal dune area in The Netherlands. Precipitation in the dunes provided and still provides fresh recharge water entering the groundwater system. The saline groundwater originally present in the system was flushed by fresh recharge water. A lens with fresh groundwater also containing a mixed water type was formed. The mixed water type in the section is labelled, F-NaHCO$_3$, and is due to cation exchange of calcium (Ca^{2+}) from a CaHCO$_3$ type of water with Na-Clay, a remnant of the more saline conditions (Stuyfzand, 1989).

5.2   GROUNDWATER CHEMISTRY AND ROCK TYPES

5.2.1   *Groundwater chemistry in consolidated rocks*

Metamorphic, intrusive and volcanic rocks

Groundwater in metamorphic, intrusive and volcanic rocks may be similar in chemical composition. Although this constituent may not be the main chemical component, dissolved silica (H_4SiO_4) is typical for the groundwater in these rock types and its behaviour has been thoroughly studied in past and present times. One may think, at first, that the mineral quartz (SiO_2), either in crystalline or amorphous form would be the

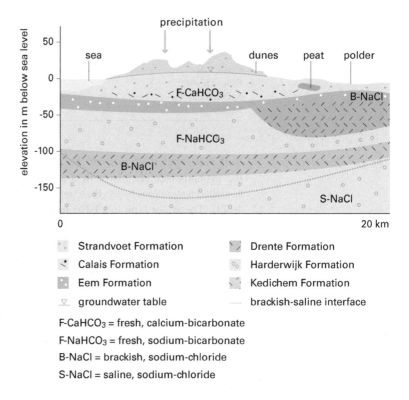

Figure 5.9. Cross section showing the hydrochemistry of a coastal area. The blue dotted line indicates the boundary between brackish and saline groundwater, Bergen, The Netherlands.

Strandvoet Formation   Drente Formation

Calais Formation   Harderwijk Formation

Eem Formation   Kedichem Formation

▽ groundwater table   ····· brackish-saline interface

F-CaHCO₃ = fresh, calcium-bicarbonate

F-NaHCO₃ = fresh, sodium-bicarbonate

B-NaCl = brackish, sodium-chloride

S-NaCl = saline, sodium-chloride

main contributor of silica to groundwater. There certainly is a contribution from quartz, but the presence of silicate minerals including feldspars, micas, and pyroxenes is more important.

The weathering of the silicate minerals forms the main driving force behind the release of silica. This weathering process usually includes dissolution dominated by hydrolysis (see section 5.1.1). The process is enhanced in an acid environment by the presence of surplus hydrogen (H^+) ions. Table 5.7 shows the weathering of orthoclase feldspar, anorthite and biotite, indicating the stimulating effect of hydrogen. In this process, clay minerals in the form of kaolinite are released and potassium (K^+), calcium (Ca^{2+}) and magnesium (Mg^{2+}) ions as well as silica are brought into solution. Once the cations and silica are dissolved, they start to take part in the flow of groundwater through the rock system.

Table 5.8 shows typical chemical compositions of groundwater in intrusive and volcanic rocks. The results are based on analyses carried out in water samples taken at wells, at springs, and at streams fed by groundwater discharge. The table indicates that groundwater in these rocks is not highly mineralised; with silica (expressed as SiO_2), bicar-

Table 5.7. Reactions showing the weathering of orthoclase-feldspar, anorthite, and biotite.

$$2KAlSi_3O_8 + 2H^+ + 9H_2O \rightarrow Al_2Si_2O_5(OH)_4 + 2K^+ + 4H_4SiO_4$$
Orthoclase feldspar $\qquad$ Kaolinite

$$CaAl_2Si_2O_8 + 2H^+ + H_2O \rightarrow Al_2Si_2O_5(OH)_4 + Ca^{2+}$$
Anorthite $\qquad$ Kaolinite

$$2K[Mg_2Fe][AlSi_3]O_{10}(OH)_2 + 8H^+ + 7H_2O \rightarrow$$
Biotite

$$Al_2Si_2O_5(OH)_4 + 2K^+ + 4Mg^{2+} + 2\,Fe(OH)_3 + 4H_4SiO_4$$
Kaolinite

Table 5.8. Groundwater composition in intrusive and volcanic rock areas in mg/l (various literature sources).

Location	pH	HCO_3^-	Cl^-	SO_4^{2-}	SiO_2	Na^+	K^+	Ca^{2+}	Mg^{2+}
Vosges/France	6.1	15.9	3.4	10.9	11.5	3.3	1.2	5.8	2.4
Massif Central/France	7.7	12.2	2.6	3.7	15.1	4.2	1.2	4.6	1.3
Senegal	7.1	43.9	4.2	0.8	46.2	8.4	2.2	8.3	3.7
Ivory Coast	5.5	6.1	< 3	0.4	10.8	0.8	1	1	0.1
Kenora/Canada	6.3	24	0.6	1.1	18.7	2.07	.59	4.8	1.54

bonate (HCO_3^-), sodium (Na^+) and calcium (Ca^{2+}) being the dominant components. The prominent place of the silica is clearly shown and competes 'for first place' with bicarbonate and sodium. The relative abundance of bicarbonate is not that surprising. The dissociation of carbonic acid and the presence of carbonate minerals in intrusive and volcanic rocks cause the release of significant amounts of bicarbonates.

Carbonate rock areas Groundwater in sedimentary rocks including sandstones, shales and carbonate rocks varies in chemical composition. With respect to chemical evolution, most intriguing are the carbonate rocks including limestones and dolomites. Typically, the main components in groundwater contained in these rocks are calcium, magnesium and bicarbonate.

Dissolution in an acid environment is the main cause for the release of calcium (Ca^{2+}), magnesium (Mg^{2+}) and bicarbonate (HCO_3^-) ions in carbonate groundwater systems (see section 5.1.1). Figure 5.10 shows diagrams describing the so-called 'open and closed system dissolution processes' in carbonate rocks. The diagrams describe the evolution paths of carbonate solution assuming typical partial pressures for carbon dioxide in groundwater. The open system process mainly occurs in the unsaturated zone and in 'open' zones in the saturated groundwater system. The carbon dioxide is usually made available by the decay of organic material. In

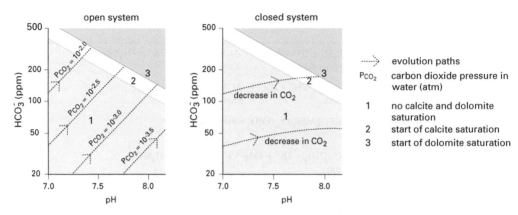

Figure 5.10. Solution in open and closed carbonate systems (modified after Langmuir, 1971).

the open system where carbon dioxide is permanently available, the dissolution of carbonates takes place until saturation is attained.

The closed system process is typical for the deeper saturated parts of the groundwater system, where dissolution takes place in groundwater flowing through opened-up fractures, and solution holes. In a closed system, the carbon dioxide available for dissolution will eventually be consumed along an evolution path (flowpath) and carbonate dissolution will virtually stop. The saturation point for calcite or dolomite may not be reached.

The dissolution of carbonate rocks in a karstification process may lead to the formation of large solution holes and eventually to caves (see also section 2.2). The orientation of caves may be correlated with the direction of bedding plane contacts, and even more so with principal zones of fracturing or weakness in the rock mass. Although the formation of solution holes and caves takes place below the groundwater table, later uplift of the rocks may expose these openings to the atmosphere.

Figure 5.11 shows solution holes in karstic limestones in the Dead Sea area in the Middle East. These solution holes which are visible at the bedding plane contacts were developed below the groundwater table, before rock uplift in this area. However, the size of the solution holes has been further enlarged as a result of dissolution by the percolation of recent precipitation.

Groundwater wich enters into caves may cause the precipitation of calcite. Well-known are the various shapes of these precipitates in caves all around the world. Precipitates in caves are called 'stalactites' in case they protrude from the ceiling of a cave, and they are referred to as 'stalagmites' where they built up from the bottom. The reason for the formation of precipitates are differences in the partial pressures of carbon dioxide. The partial pressure of carbon dioxide in saturated carbonate rocks is usually higher than the partial pressure of carbon dioxide in caves where atmospheric conditions prevail. Groundwater entering a

Figure 5.11. Karstic limestones with solution holes at bedding plane contacts, Jericho in the Dead Sea area.

Table 5.9. Groundwater composition in carbonate rocks in mg/l (various literature sources).

Location/Description	Temp (°C)	pH	HCO_3^-	Cl^-	SO_4^{2-}	Na^+	K^+	Ca^{2+}	Mg^{2+}
Manitoba/Silurian dolomite	5	7.6	417	27	96	37	9	60	60
Pennsylvania/Limestone and dolomite	11	7.4	183	8	22	4	1.6	48	14
Palestine/Limestone	>15	-	218	30	12	18	2.3	52	21
Florida/Limestone	24	7.7	160	12	53	8	1	56	12

cave with a lower carbon dioxide pressure will trigger off the precipitation of calcite (see also equation (5.9) in section 5.1).

Table 5.9 shows the chemical compositions of groundwater in typical carbonate rock areas. These compositions are based on the analyses of water samples collected in wells in Manitoba and Florida and at springs in Pennsylvania and the Middle East. The relative abundance of calcium (Ca^{2+}), magnesium (Mg^{2+}) and carbonate (HCO_3^-) ions is clearly shown. The Manitoba dolomites have the highest concentrations of dissolved minerals. The high concentrations can be attributed to the abundance of carbon dioxide in the relatively cold groundwater, originating from a surplus of organic material.

The total concentrations of dissolved minerals in carbonate rocks are large in comparison with the total concentrations in groundwater in metamorphic, intrusive and volcanic rocks. Apparently, the dissolution in these latter rocks is less effective than in carbonate rocks. The higher mineralisation of groundwater in carbonate rocks can be explained by the higher solubilities of the individual minerals: calcite and dolomite. Note, however, that carbonate rocks usually have much smaller silica concentrations than metamorphic, intrusive or volcanic rocks. If present at all in carbonate rocks, then the silica originates from chert bands and other silicate minerals, locally present in these rocks.

5.2.2   *Groundwater chemistry in unconsolidated rocks*

Unconsolidated sediments   The chemistry of groundwater contained in unconsolidated rock can be extremely diverse. As opposed to the rather monotonous carbonate rocks, unconsolidated rocks including gravels, sands, silts, clays, peats, and halites have more variation in mineral content. The main components in groundwater contained in unconsolidated sediments include calcium, magnesium, sodium, bicarbonate and chloride.

In unconsolidated rocks, dissolution processes play an important role in the release of chemical constituents in groundwater. Carbon dioxide loaden recharge water will not readily dissolve the quartz in a gravel or sand, but in case calcite is contained in these rocks, the release of calcium (Ca^{2+}) and bicarbonate (HCO_3^-) ions takes place. Groundwater flowing through gypsum ($CaSO_4$) deposits even more effectively dissolves this mineral and produces calcium and sulphate (SO_4^{2-}) ions in large quantities. Redox reactions, ion exchange, mixing, flushing and precipitation are also essential processes taking place in unconsolidated sedimentary systems. Redox reactions take place in all sorts of unconsolidated sediments and involve the transformation of nitrates, sulphates, sulphides, oxides, and arsenates. In sediments, ion exchange is typical for the silty and in particular the clayey and peaty parts of the deposits.

In view of its large variation, the determination of groundwater chemistry in unconsolidated rocks is an elaborate, but essential activity. Water samples usually have to be collected in wells at various locations. Figure 5.12 shows the typical collection of a water sample used for the determination of major constituents. The water sample is taken at a well over 50 m in depth and covered by pavement. The device shown in the picture consists of a stainless steel hollow tube with a valve at the bottom that traps the groundwater as soon as the tube has submerged into the water.

Table 5.10 shows the chemical components of groundwater representative for unconsolidated sediments of various origins. The shown chemical compositions reflect the processes that have influenced the

Figure 5.12. Water sampling with a simple device at a covered well in Limburg, The Netherlands.

Table 5.10. Groundwater composition in unconsolidated sediments in mg/l (various literature sources).

Location/Description	pH	HCO_3^-	Cl^-	SO_4^{2-}	Na^+	K^+	Ca^{2+}	Mg^{2+}
Netherlands/Sandy fluviatile deposits [1]	7.7	201	10	3	8	0.9	54	7.7
Netherlands/Sandy fluviatile deposits [2]	7.6	322	49	7	36	6.2	76	12
Egypt/Sandy fluviatile deposits [3]	7.4	114	19	21	35	4	9.6	11
Australia/Sandy eolian deposits [4]	6.7	100	110	32	52	6	27	8.8

[1] Inland area.
[2] Coastal area, overlain by marine fine sandy/clayey deposits.
[3] Overlain by clays.
[4] Coastal area, overlain by minor peat and marine clays.

groundwater chemistry of the sediments. The first sample collected in coarse sandy fluviatile deposits in the eastern part of The Netherlands reflects the high amount of calcite ($CaCO_3$) contained in the rock. Dissolution of the calcite has led to the relatively high calcium (Ca^{2+}) and bicarbonate (HCO$_3^-$) concentrations. The second sample taken in the coastal area of the western part of The Netherlands was drawn from a similar rock type. The calcium and bicarbonate concentrations are also high, but elevated amounts of sodium (Na$^+$) and chloride (Cl$^-$) ions are also shown. The sodium and chloride originate from recharge by precipitation which, in a coastal area, is known to have elevated concentrations of these elements. However, the high sodium and chloride concentrations may also be attributed to the flushing of these ions from overlying marine deposits into the sampled sandy fluviatile sediments.

The third sample, obtained from medium to coarse sandy deposits at the margin of the Nile delta in Egypt shows a relative high concentration of sodium (Na$^+$) ions in relation to both chloride (Cl$^-$) and calcium (Ca^{2+}) ions. The explanation that may be offered relates to the presence of clays above the sampled sands. Ion exchange has removed calcium from down-flowing groundwater and replaced this component by sodium. The fourth sample has been collected in eolian sands near the Australian coastline in Sydney. Higher up in the sequence marine clay beds, peat and shell fragments have been identified. The high sodium (Na$^+$) and chloride (Cl$^-$) ion concentrations point to continued flushing of the eolian sands that were previously inundated with saline water. The elevated calcium (Ca^{2+}) concentration can be attributed to groundwater flowing first through the calcite-rich shell layer, before it enters the eolian sands.

CHAPTER 6

Development of Groundwater

6.1 GROUNDWATER MANAGEMENT

6.1.1 *Water management*

Introduction

Water management can be defined as 'the pre-occupation of decision-makers and professionals to develop, control and protect the earth's water resources for the benefit of mankind'. Therefore, staff involved in the management of water resources may have one of the following main objectives in mind:
− Development of water resources
− Control of hazards
− Protection of the environment

To satisfy these objectives, numerous activities may be carried out. Table 6.1 shows a brief selection of water-related activities. For the development of water resources, for example, the design and construction of diversion weirs, water intakes and distribution networks may be undertaken to obtain water supplies from rivers. The activities may also cover the design and installation of wells or well fields to secure groundwater supplies. The control of hazards may concern the planning and the construction of dikes along a river or coast for flood protection or the installation of pipe drains to lower groundwater tables in agricultural land in order to improve drainage conditions. Protection of the environment covers a wide range of activities including the implementation of measures to prevent contaminants from households, industries or agriculture entering water resources systems.

Figure 6.1 presents a historical example of water resources development in Yemen. The picture shows one of the intake channels at the ancient dam of the Queen of Sheba. The dam was constructed to set up the flood water levels in the wadi Adhana. In this way floodwater could flow under gravity to agricultural land for irrigation purposes. Unfortunately, the irrigation scheme came to an end when the silt load collected behind the dam caused the collapse of this structure.

One clearly cannot pursue one of the main objectives outlined above without giving consideration to the other objectives. The installation of a

Table 6.1. Water management objectives and selection of related activities.

Objective	Water source	Activities
Development of water resources	Surface water	Building of diversion weirs and intakes, construction of storage dams
	Groundwater	Construction of spring captations, installation of well fields, implementation of artificial recharge schemes
Control of hazards	Surface water	Setting up of dikes against floods, widening and deepening of river channels for flood storage
	Groundwater	Construction of pipe drains and ditches, installation of pump systems for mine and excavation dewatering schemes
Protection of the environment	Surface water	Clean-up of spills in river systems, watershed management to prevent catchment contamination
	Groundwater	Installation of proper sealing at waste disposal and water treatment sites to prevent household and industrial contamination, installation of 'pump and treat' systems to assist in the clean-up of groundwater contamination

Figure 6.1. Remains of the Queen of Sheba Dam, Marib, Yemen.

water supply to develop groundwater resources may be in conflict with the control of hazards or with the protection of the environment. Consider an example. Imagine that a city wants to develop the local groundwater resources by installing wells in a coastal area. The city also wants to control hazards resulting from land subsidence in the area. Abstracting groundwater at wells may conflict with the prevention of land subsidence. Decision-makers and professionals will have to formulate a groundwater development scenario whereby land subsidence is minimised.

Strategies for water resources development

Water management issues concerning the development of water resources are discussed in more detail in this textbook. Water resources can only be successfully developed when sound strategies are formulated. To set out a strategy for the development of water resources on a country wide, provincial or local scale, water related factors should be considered. These factors concern the water resources system, the demand for water, the environmental impact, and the socio-economy.

Figure 6.2 presents a diagram showing the interaction between these factors. How does this diagram work? The factor 'water resource'

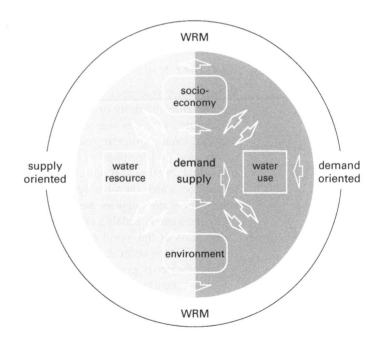

Figure 6.2. The interaction between factors for water resources development (modified after Koudstaal et al., 1992).

hydro(geo)logy

water engineering

WRM   water resources management

indicates the physical water resources system itself and 'water use' concerns the water that is demanded and used for all sorts of purposes. In areas with water shortages actions may be undertaken towards further development of the water resources system. In case one is merely looking at satisfying all demands on water utilising the water resources system, a 'supply-oriented' approach is followed. However, one may also restrict the demand for water. In that case, reference is made to a 'demand-oriented' approach.

The factor 'environment' refers to the changes in the natural environment caused by the installation of water supplies that tap from a particular water resources system. These changes may not be desired, so that either the demand can be restricted following the demand-oriented approach or another water resources system is developed. 'Socio-economic' factors including the costs and benefits of the related water supply schemes are also evaluated. In case costs and social burdens are too high then also the demand could be reduced or another system is selected for development. Careful weighing between supply and demand oriented approaches, and the evaluation of environmental and socio-economic factors is a responsible task, which should lead to the formulation of an optimum water development strategy.

6.1.2 *The role of groundwater*

Advantages and disadvantages of groundwater

Which water resources can be developed to secure a water supply for a small community or town, for industrial complexes or irrigated agriculture? One can use groundwater resources, surface water, or one can even exploit resources including seawater that has to be desalinated or precipitation (rain) that is collected on the roofs of individual dwellings. If one concentrates on groundwater as opposed to surface water then the typical advantages of groundwater as a source for water supplies are:

– Groundwater is widely available. In particular in many semi-arid and arid parts of the world groundwater can be developed at locations where surface water is not exploitable.
– Groundwater is available throughout the year. A groundwater system is a nearly perfect storage medium for water. In dry periods when surface water supplies dry up, water can still be supplied from groundwater resources.
– Groundwater quality is superior. In case a groundwater system can be identified then the quality of its water is usually superior to the quality of surface water. Groundwater does not contain sediment and suspended material, neither has a bad smell, taste or coloring. The temperature is relatively low and constant throughout the year. Groundwater is normally hygienically safe.

A supply from groundwater may also have a few drawbacks. The most relevant shortcomings can be listed as follows:
- Vast supplies of groundwater are not everywhere available. Water demands may then have to be met from surface water. For instance, most of the very large cities on earth are (partly) supplied by surface water.
- Installations that supply groundwater may be prone to failure. In many places small diameter wells supply groundwater into distribution systems. Although they are generally quite dependable, these wells could, without proper operation and maintenance, suffer from breakdowns.

Cost considerations

The advantages and disadvantages outlined above relate to the physical characteristics of water resources systems. How does the development of groundwater systems or surface water resources relate to costs? The expenses may include investment, and operation and maintenance costs of the supply systems. The costs will have to be recovered through government subsidies and contributions from individual families, farmers and industries.

Table 6.2 shows a tentative summary of the cost level for the different sources that can be used for water supply schemes. It can be concluded from the table that supplies from groundwater are relatively inexpensive. The table also shows the advantage of the development of natural springs. If they are located nearby water users, pumping is not required, and sound hygienic protection can be implemented, then springs are the most ideal sources for water supply schemes. If surface water resources are implemented then the use of this source for drinking water supplies is quite costly when pumping and extensive treatment is required.

Table 6.2. Cost levels of water supply sources.

Source	Supply scheme	Level investment costs	Level operation and maintenance costs
Groundwater	Springs	Low	Low
	Dug wells	Low	Low
	Shallow tubewells	Low/medium	Low/medium
	Deep tubewells	Medium	Medium
Surface water	Pumped/chemical treatment	Medium/high	High
	Pumped/slow sand filtration	Medium/high	Medium/high
Seawater	Desalination plant	Very high	Very high
Precipitation	Dams/reservoirs	Medium/high	Medium
	Cisterns	Medium/high	Low

6.2 GROUNDWATER PLANNING AND INVESTIGATIONS

6.2.1 *Water planning*

Regional masterplans
and local planning

Masterplans outlining the general development of water resources are prepared for an increasing number of areas in the world. The plans take into account the characteristics of- and the interrelationships between the various water resources development factors discussed in section 6.1.2. First of all, these plans include study objectives, focusing on:
– Identification of water resources systems
– Determination of the extent of these systems
– Assessment of the amounts of available water
– Evaluation of water quality

Masterplans also provide estimates on present water use and future water demand. In addition they point out the environmental consequences of water resources development and they give information on socio-economic aspects. Masterplans are usually prepared on a regional scale. They may be drawn up for a country, a group of countries or a province, for an entire river catchment, or for a groundwater system.

Once masterplans have been drawn up, water resources development activities may proceed on a local scale. Figure 6.3 shows a diagram illustrating that the local activities may follow a cyclic behaviour. Within a cycle, a number of phased activities can usually be distinguished, which are: 1) study (investigation) activities, 2) design activities, 3) installation of infrastructure, and 4) continuation of a monitoring programme. Obviously, the studies come up first. An often formulated study objective concerns the selection of sites for water resources development, taking into account the local dimensions of the source and the related water availability and quality, local water demand and other factors.

Based on the study results, the design activities for the water supply infrastructure are taken at hand. In turn, the designs form the basis for the installation of infrastructure. The monitoring programme refers to activities that are set up to check the effects of the installation of a water supply. A proper monitoring program is already set up before the construction of the infrastructure is completed and should continue during the succeeding water resources development cycles.

Two illustrative examples of regional water planning and local water resources development are presented. Within the context of a provincial masterplan, the use of surface water from a river system, identified in the province, is recommended. Groundwater plays a role, but is less important. Estimates on the extent of the river system and the amounts of water that are available are outlined in the masterplan studies. At a local scale, studies and designs are completed for a phased setting up

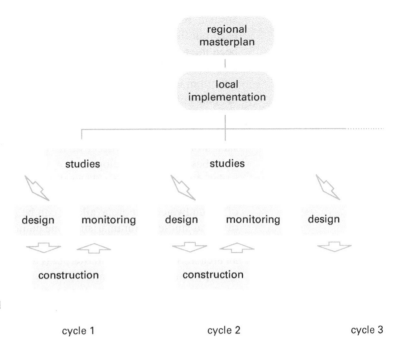

Figure 6.3. Regional and local activities (cycle 3 only partly shown).

of intake structures. On the basis of local geological studies and river discharge measurements, suitable sites for intake structures are selected during various project cycles. Upon completion of the designs, intake structures, distribution networks and monitoring facilities are put into place.

Consider also an example whereby the focus is more on the use of groundwater than on surface water. Imagine that studies carried out for the formulation of a regional masterplan indicate that a basin is underlain by a groundwater system consisting of a vast sandstone aquifer. Regional studies also form the basis for estimates on the extent of the system and the available amounts of groundwater that can be taken from these sandstones. Local studies and designs for the installation of infrastructure during a series of project cycles are carried out. The studies reveal suitable sites for wells and provide insight into the local amounts of water that can be taken for abstraction. Finally, the infrastructure is installed and groundwater monitoring can proceed to establish the effects of the water supply schemes.

'Top-down' and 'bottom up' approach

To start with masterplans and to follow these plans up with local studies and the implementation of water supply schemes is an interesting 'top-down' approach. However, reality has usually been the other way around. For example, without proper masterplans many wells have been put into place. Such an approach without masterplans may have been the result of a lack of government laws enforcing these plans or a lack of funds for the completion of regional studies. Although such a 'bottom up' approach may be quite disappointing, it is noted that on the basis of the information that has been collected at the local water supply schemes, masterplans may still be drawn up.

In addition, the formulation of masterplans is not always necessary. For example, consider areas where water supply schemes have to be based on the development of small groundwater systems that are isolated from each other. Then the preparation of masterplans is less pressing and issues regarding the extent of the systems and the amounts of groundwater that are available can be addressed within the context of 'local studies'. Summarising, one could say that working with masterplans should be considered at all times. However, the conditions in an area may be such that masterplans are not strictly required and only local planning is sufficient.

6.2.2 *Selection of investigation methods*

General outline

For the development of groundwater resources, the groundwater system needs to be carefully studied (see section 6.2.1). These studies usually include a number of field investigations. The most popular investigation methods are described in sections 2.3, 3.2.4 and 4.2.

The selection of a proper set of investigation methods depends on the study objectives that have been formulated. Table 6.3 has been prepared to indicate common objectives for groundwater studies and the related investigation methods that can be selected. The table does not restrict itself to water resources development, but also takes into account the 'water areas', control of hazards and protection of the environment (see section 6.1.1).

An illustrative example of the use of Table 6.3 can be considered. Assume that a regional groundwater system has been identified and its extent has been determined through the engagement of investigations as listed in the table. Imagine that one of the study objectives further concerns the assessment of the amount of available groundwater resources. To satisfy the objective, one focuses on setting up groundwater balances (see section 6.3.1). This includes the application of groundwater balance estimation methods. Additional investigations that may be set up comprise the collection of meteorological data.

Table 6.3 does not pretend to give a complete guideline for the selection of investigation methods, nor rules that should be followed at all times. In particular, the selection of field investigation methods also depends on the complexity of the area, the data that

Table 6.3. Selection of investigation methods.

Study type	Study objective	Approach	Investigation method [1]
Water resources development	Identification of groundwater systems	Hydrogeological classification	A,B,C
	-----	-----	-----
	Determination of the (lateral and vertical) extent of the system	Dimensioning of aquifers and aquitards	(A,B,C),D,E,F
	-----	-----	-----
	Assessment of available groundwater	Groundwater balance compilation [2]	(E),F,G
	-----	-----	-----
	Selection of abstraction sites (e.g. well or well field)	Local dimensioning of aquifers and aquitards Local water balances [2]	A,B,C,D,E,F,G
Control of hazards	Groundwater drainage: amount of water to be drained	Dimensioning of aquifers and aquitards Water balance compilation	A,B,C,D,E,F,G
Protection of the environment	Waste disposal: distribution of contaminant plumes	Dimensioning of aquifers and aquitards and water quality assessment	B,C,D,E

[1] Desk study of available information is always carried out.
Methods described in section 2.3
A: Satellite imagery and aerial photography.
B: Hydrogeological mapping.
C: Well inventories.
D: Geophysical surface investigations.
E: Exploration or test hole drilling and logging.
Method outlined in 3.2.4
F: Pumping tests.
Methods discussed in section 4.2
G: Groundwater balance estimation methods.
[2] Meteorological data may also be needed for groundwater balance estimates.

are available and the financial resources that can be used to collect new information.

Figure 6.4 shows a 'flow diagram' that has been prepared to illustrate the 'placement' of selected investigation methods in the correct time frame. The case portrayed in the flow diagram concerned studies carried out in the Quetta region in west Pakistan (Euroconsult, 1988). The water resources development studies for the region focused on the supply of groundwater from scattered alluvial

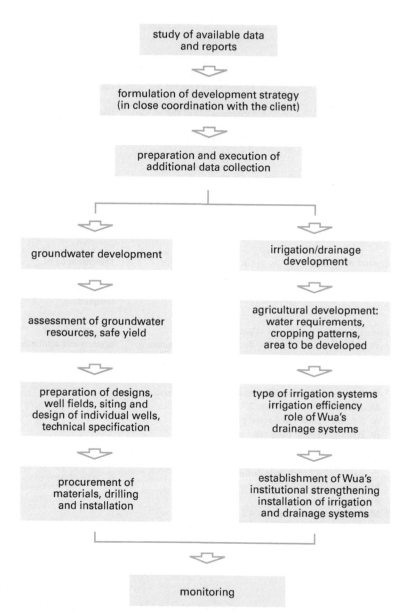

Figure 6.4. Flow diagram
of activities for groundwater
projects in the Quetta region,
Pakistan.

groundwater systems. The study objectives comprised the identification of groundwater systems, the determination of the extent of these systems, and the assessment of the available groundwater resources. Another objective entailed the local selection of well sites and the completion of well design.

The flow diagram shows that the first step comprised a desk study of the available data and reports followed by the formulation of a development strategy and the collection of additional data, mainly through field investigations. The investigations to determine the extent of the groundwater systems included aerial photography, hydrogeological mapping, well inventories, geophysical surveys and exploration drilling. Then, in separate steps, groundwater assessments which included the determination of available water on the basis of groundwater balance estimation methods, and irrigation evaluations were completed. To be further introduced in section 6.2.3, monitoring activities formed the last step in the flow diagram.

6.2.3 Groundwater monitoring

Monitoring practises

Groundwater monitoring may be defined as the 'recording of time series of groundwater data to follow up the effects of the implementation of infrastructural works including the captation of springs, the construction of wells and well fields, and even the installation of artificial recharge'. Measurement periods for monitoring may be in the order of several years and even longer. Observations may be taken at hourly, daily, monthly or even yearly intervals. Continuous registration of data can also be carried out. Parameters that are normally measured in monitoring wells include groundwater levels, groundwater abstractions, and data on groundwater quality.

Figure 6.5 shows the result of a monitoring programme that was set up to assess the impact of a well field constructed for the city of Zhengzhou in China (MGMR & TNO, 1989). The local groundwater

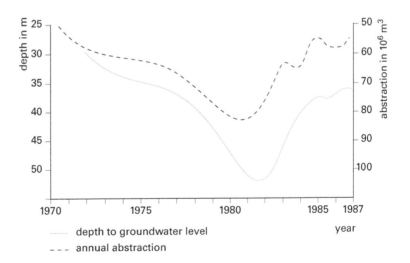

Figure 6.5. Graphs of groundwater levels and abstraction rates in the Zhengzhou area, China.

—— depth to groundwater level

- - - annual abstraction

system, introduced in section 3.2.2, consists of a shallow and deep aquifer. The well field taps the deep aquifer. Groundwater levels were recorded in an observation well, monitoring this aquifer. The figure shows the impact of the well field, illustrated in a graph of groundwater levels (as depth to groundwater level) plotted against well abstractions. The continuing and significant downward trend in the groundwater levels up to 1982 along with the increase in abstractions pointed to a severe reduction in groundwater storage (see section 6.3.1). However, the reduction in the abstraction rates since 1982 raised the groundwater levels significantly and (partly) restored the groundwater system in the Zhengzhou area.

Groundwater monitoring programmes to assess 'the effect of contamination from domestic, agricultural or industrial activities on the groundwater system' may also be set up. The engagement of these programmes has rapidly increased in recent years. The release of

Figure 6.6. Groundwater level monitoring at a landfill site, Bennekom, The Netherlands.

contaminants may severely affect the system and endanger a well or well field tapping from the system. Figure 6.6 illustrates the monitoring of groundwater levels in an observation well near a landfill site located in the central part of The Netherlands. Water samples for the analyses of groundwater quality are also recovered at the well. The landfill is located on top of a coarse sandy and gravelly aquifer overlain by finer material. The measurements may lead to the detection of contaminants leaking from the landfill site into the groundwater system which is being exploited at a nearby well field.

6.3 GROUNDWATER RESOURCES STUDIES

6.3.1 *Regional groundwater availability*

Long and short term availability

For discussion in this section, a most intriguing type of groundwater study has been selected. This concerns the groundwater resources development study, which focuses on the assessment of available water resources. In principle, the study covers the assessment of the amounts of groundwater that can be made available from regional groundwater systems that have been identified for groundwater development (see section 6.2.1 and Table 6.3). Assume that the lateral and vertical extents of these systems have been properly determined. Groundwater availability assessments for the long term and the short term can then be carried out, following a specific strategy.

Two strategies are traditionally considered when one makes assessments on groundwater availabilities for the long term, usually covering many years. One strategy is that 'in the long run' the groundwater made available for abstractions does not continually reduce the amount of groundwater stored in the groundwater system. This strategy is also called the 'safe yield' approach. In a somewhat broader way safe yield can also be defined as 'the abstractions that can be maintained over a long period of time without causing an unacceptable reduction in groundwater storage and decrease in groundwater tables, an unacceptable pumping lift, or initiating a decline in groundwater quality'. Another strategy is to allow a serious reduction in the amount of groundwater stored in the system. This approach carries the risk that the groundwater resources will be depleted. The generally accepted opinion is that the first strategy should be followed.

How could one approach the groundwater availability assessments for the short term? Imagine that periods of a few months or even a few consecutive years are considered. A generally accepted strategy for short periods is that part of the amount of water stored in the groundwater system can be used. Obviously, the strategy is only viable when the amounts of groundwater stored are sufficient and groundwater quality does not detoriate. This strategy caters in particular for the abstraction of groundwater in dry seasons or during years with below average precipitation. The underlying assumption is that in the wet seasons and in years with above average precipitation the groundwater stored in the system is brought back to average levels.

Methodology for groundwater availability assessments

Various methods are in use to assess groundwater availability. Table 6.4 presents a summary of the traditional methods that are in use to determine groundwater availability on the basis of groundwater recharge assessments. These recharges are usually estimated by considering one of the components of the groundwater balance. There is not that much

Table 6.4. Methods to determine aquifer recharge.

Selected methods	Brief description	Comment
1. *Hydrometeorological techniques*		
Water balance method	Recharge into aquifer is computed from precipitation, evapotranspiration, runoff and soil moisture data	Evapotranspiration is difficult to determine
River/spring hydrograph analysis	Baseflow (for rivers) component of hydrograph is determined which relates to groundwater discharge	Groundwater discharge from baseflows may be hard to equate to total groundwater recharge
2. *Hydrogeological techniques*		
Groundwater table fluctuation method	Seasonal rises in groundwater table are converted into recharge	Specific yield of the aquifer may be difficult to determine (especially in hard rock areas)
Flownet analysis	Flownets of hydraulic heads are combined with Darcy's Law to compute aquifer throughflow	Effective method in not too complex and undisturbed aquifers
3. *Hydrogeochemical techniques*		
Chloride method	Chloride concentrations in precipitation and groundwater are compared to compute aquifer recharge	Chlorides in groundwater from other sources than precipitation may lead to less accurate estimates
Isotope methods	For example, oxygen isotopes distributions in groundwater are determined to assess recharge	Effective method, but requires advanced groundwater sampling techniques and estimation of effective porosity
4. *Composite techniques*		
Groundwater balance method [1]	Measurable and determinable recharge and discharge components are compared in a water balance	Effective method considering all recharge terms, with cross checking, but may be tedious
Groundwater modelling	Well-calibrated groundwater models provide comprehensive groundwater balances including recharge	Non-uniqueness of model solutions may lead to erroneous values of recharge

[1] Method selected for discussion in this textbook.

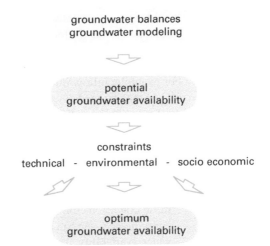

groundwater balances
groundwater modeling

potential
groundwater availability

constraints
technical - environmental - socio economic

optimum
groundwater availability

Figure 6.7. Diagram showing
groundwater availability
assessment.

emphasis on the determination of the complete balance (see section 4.1). In this book, however, the complete balance will be considered for the determination of groundwater availability. The risk of ignoring one of the recharge, or other inflow terms, is then minimal.

Based on the outlined strategies, the approach that is proposed for the determination of long term and short term availability includes assessments on the 'potential' and 'optimum' available amounts of groundwater. The basic idea underlying this approach to assess groundwater availability then is, that abstractions for groundwater development should not exceed the 'optimum' available amounts. Figure 6.7 shows a diagram presenting the overall approach, while further details could be outlined as follows:

i) *Long term availability.* Consider the groundwater development strategy for the long term, i.e. the groundwater available for abstraction should (after some initial storage reduction) not cause a continuous decline of the amount of groundwater stored in the groundwater system. This strategy has implications for the groundwater balance. The right hand side of the general groundwater balance equation as formulated in section 4.1 is limited to zero; in other words one is dealing with the equilibrium balance:

$$I - O = 0 \qquad\qquad (6.1)$$

where:
I = total rate of groundwater inflow (m^3/day)
O = total rate of groundwater outflow (m^3/day)

The potential groundwater availability can now be defined as the 'total long term groundwater inflow into the groundwater system'. The assumption is that the total groundwater inflow (I) is potentially available for abstraction and this will be at the expense of the natural groundwater outflow (O). In that case groundwater storage will remain intact. For example, inflow into a system such as recharge from precipitation (Q_{prec}) may be made available at the expense of an outflow like spring discharge (Q_{spring}). When using the above definition for the potential amount of available groundwater one also assumes that:

– The inflow components will not change when abstractions are put into place. This is in many places, but not everywhere, the case. In section 4.1.2 it is pointed out that, for example, the recharge from precipitation may increase when abstractions are implemented. For those conditions the total inflow into the system increases and, as a consequence, one can also set a larger value for the potential groundwater availability.

– There is no future augmentation of the inflow into the groundwater system by artificial means. For the case that, at some stage, artificial recharge schemes are implemented, the recharge amounts (Q_{art}) should be added to the potential groundwater availability.

– The water quality in the groundwater system is acceptable. For the case that part of the groundwater inflow (or outflow) has an unacceptable quality, the potential groundwater availability would be less than these quantities. The potential availability can then be assessed by subtraction of the part with the poor water quality.

The optimum groundwater availability is derived from the potential amount of available groundwater. The optimum availability can be considered as the 'groundwater availability taking into account technical, environmental and socio-economic constraints'. Since the constraints reduce the amounts of groundwater available for abstraction, the optimum availability is less than the potential availability. The relations can be described as follows:

– *Technical constraints:* These constraints include the inability of abstraction wells to intercept all the groundwater before it flows out of the system. For example, the permeabilities of the rocks can be so low that even a large number of wells will not be capable to pump up all the groundwater before it discharges from the system in the form of capillary flow (Q_{cap}), springflow (Q_{spring}), or discharge to streams and rivers ($Q_{surfout}$).

– *Environmental constraints:* These constraints concern the damage that abstractions from wells may have on the natural environment in an area. For example, decreased groundwater levels in well field areas may affect the amount of water required for consumption by natural vegetation. Also, lowered groundwater levels may alter groundwater flow directions that may deprive the vegetation of essential nutrients. In addition, severe damage to the natural environment is done when a groundwater system turns saline as a result of excessive abstractions and the termination of groundwater outflow. This outflow will have to be maintained when inflows into the system contain elevated concentrations of salts: these salts will have to be exported.

– *Socio-economic constraints:* These constraints focus on the effects that abstractions may have on the socio-economic framework of an area. For example, springs existing in an area may have long been used to secure a supply of water for domestic and agricultural purposes. When groundwater abstractions from wells are implemented, springs may dry up and social unrest may follow. Therefore, existing groundwater supplies will have to be secured.

Economic factors are usually related to costs. The cost of groundwater abstraction by wells depends on the depth and yield of the individual wells. When are these costs too high and prohibit further development of groundwater resources? This depends on the groundwater development objectives and the alternatives that are at hand. For example, rather high costs for the construction and operation of wells used for domestic supplies are generally accepted if no other sources are available. One cannot do without drinking water. For the development of groundwater for irrigation other criteria are set. Cost-benefit studies are usually made: when there are insufficient benefits to justify the costs, groundwater development for irrigation will not go ahead.

For a particular groundwater system the cumulative effect of all the constraints outlined above should be weighed. This weighing allows the assessment of the optimum groundwater availability on the basis of the potential groundwater availability. There are no hard rules for this assessment. Traditionally, these assessments are done in a rather qualitative way, but more quantitative methods involving the use of groundwater models have recently gained in popularity (see also below).

The optimum groundwater availabilities for well known groundwater systems around the world have often been set somewhere in the range of 30% to 80% of the potential availabilities. For example, for the alluvial sandy and gravelly groundwater system in the eastern part of The Netherlands, the optimum availability has been set at only 30% of the potential groundwater availability. On the other hand, for the alluvial sandy system in the coastal area of Haiti, the optimum availability was considered to be 70% of the potential groundwater availability.

ii) *Short term availability.* For the short term groundwater development strategy, allowing a temporal reduction in the amount of water stored in the groundwater system during dry periods, a potentially available amount of groundwater storage may be formulated. To find a suitable expression for this storage, the right hand side of the groundwater balance equation (4.13) may be modified. In case a volume is considered instead of a volumetric rate then:

$$V_{bal} = GS_y \, (\phi_{ave} - \phi_{min}) \tag{6.2}$$

where:

V_{bal} = volume of groundwater available from storage (m^3)
G = surface area (m^2)
S_y = specific yield (dimensionless)
ϕ_{ave} = average groundwater table elevation (m)
ϕ_{min}= minimum groundwater table elevation (m)

The potentially available amount of groundwater storage in a groundwater system can then be defined as the V_{bal} in equation (6.2). The ϕ_{ave} refers to the average long-term groundwater table for equilibrium conditions, and the ϕ_{min} indicates a minimum, but recoverable groundwater table in the groundwater system. The specific yield should be representative for all the rocks in the groundwater system. It may be rather 'tricky' to determine a representative specific yield. First of all this parameter may change considerably from place to place in the area concerned. Secondly, the specific yield may also show a large variation with depth. Due to the effect of compaction the specific yield at a larger depth is usually smaller than near the surface.

The optimum amount of groundwater that can be taken from groundwater storage for the short term can be derived from the potential amount. This optimum volume of groundwater could be described as the 'volume that can be abstracted for the short term, giving due consideration to technical, environmental and socio-economical constraints'. Similar to the assessment of the optimum long term availability, the optimum available volume from storage can be determined on the basis of the potentially available groundwater storage and the 'weight' of the constraints. The methods used for this determination are also largely qualitative, but more quantitative methods using groundwater models are increasingly being used.

The role of groundwater modelling

The key to the assessment of the potential long and short term groundwater availabilities is the compilation of groundwater balances for well defined groundwater systems (see section 4.2). Groundwater flow models (section 3.2) may be engaged to check whether the compiled groundwater balances are realistic. In case a groundwater model is not able to reproduce the groundwater levels measured at observation wells, then the groundwater balance terms for an area might have been been incorrectly assessed.

Optimum groundwater availabilities can be estimated with the help of groundwater flow models, management models, mass transport models and fresh-saline interface models. The models predict the effects of abstractions on groundwater flows, groundwater levels (hydraulic heads) and groundwater quality. An abstraction scenario may be selected for implementation that minimises these effects and the related constraints on the environment and the socio-economy. Optimum groundwater availabilities correspond with the groundwater balance terms defined for the selected scenario.

Figure 6.8 shows the model-computed groundwater levels relating to one of the proposed groundwater development scenarios in the Ping Tung Plain in Taiwan (Ting, 1997). This coastal plain is underlain by a groundwater system consisting of three sandy and gravelly aquifers, separated by aquitards. The model included flow and management components and was used to define an abstraction scenario that minimised the constraints. The model-computed groundwater balance indicated a substantial long-term optimum groundwater

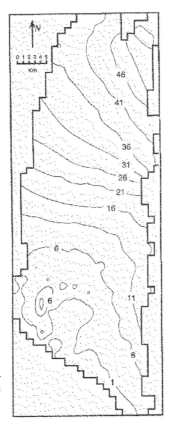

Figure 6.8. Model based map showing groundwater level contour lines (in m above mean sea level), relating to an optimum abstraction and artificial recharge scenario from the upper aquifer in the Ping Tung Plain, Taiwan.

availability in the order of $880*10^6$ m³/year. The model was also engaged to assess the effect of artificial recharge to combat saline groundwater intrusion at the coast (see below). With the artificial recharge in place, the optimum groundwater availability increased to $920*10^6$ m³/year.

Artificial recharge

Artificial recharge is defined in section 1.2.5 as 'the practice of increasing by artificial means the amount of water that is entering the groundwater system'. Infiltration ponds or basins, galleries, or injection wells can be installed to supply surface water or treated waste water into the system. Figure 6.9 shows cross sections indicating the set ups for these infiltration systems. The cross section clearly shows the infiltration structures and the wells that can be placed to abstract the recharge water.

Recharge by artificial means, introducing the term Q_{art} into the groundwater balance (see section 4.1) has a positive effect on the long and short term groundwater availabilities. These and other advantages of artificial recharge schemes can be summarised as follows:

– *Long term availability.* In case the long term inflow into a groundwater system is less than the long term demand for groundwater, then artificial recharge provides a means of increasing the groundwater availability.

– *Short term availability.* If groundwater storage in dry periods is quickly depleted, then artificial recharge can be considered. In wet periods surface water may be diverted to infiltration works and the water is stored underground for later use during dry periods. This principle can also be considered as one of the bases for so-called 'conjunctive use' schemes that aim to optimize the development of water resources of an area by adopting an integrated surface water-groundwater management policy.

– *Storage facilities.* In some areas there may be a lack of suitable sites for surface water reservoirs, or there may be strong opposition against the construction of such reservoirs. Under these circumstances the storage of surface water in a groundwater system through artificial recharge may be an attractive alternative.

– *Restoration of groundwater levels.* In many areas where groundwater abstractions are taking place, the depressed groundwater levels can be restored by the application of artificial recharge. Figure 6.10 shows a location map and a cross section through an artificial recharge gallery in the eastern part of The Netherlands. The scheme was designed to restore groundwater levels in a nature reserve area that supposedly had been affected by abstractions at a nearby well field.

– *Groundwater quality.* In case the existing groundwater quality in a groundwater system is poor as a result of the presence of fossil

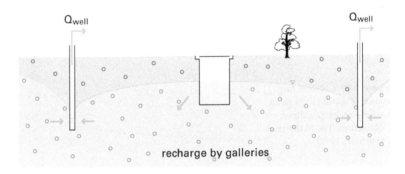

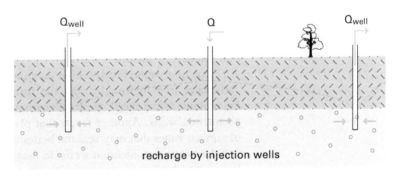

Figure 6.9. Cross sections showing various types of artificial recharge.

○	aquifer	▽	groundwater table
╱	aquitard	---	piezometric line

groundwater or saline intrusion, then artificial recharge of fresh water may have a beneficial effect. Also, the quality of infiltration water affected by pathogenic bacteria and viruses may improve when it is used in artificial recharge schemes.

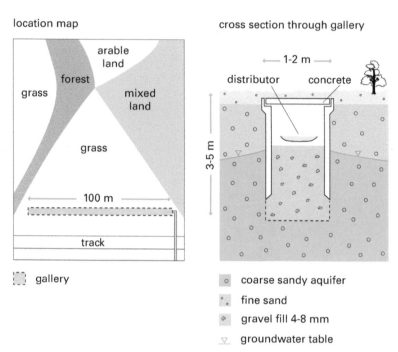

location map   cross section through gallery

Figure 6.10. Location map
and cross section through an
artificial recharge gallery,
Doetinchem, The Netherlands
(Abushaar, 1997).

There are some risks involved in the introduction of artificial recharge schemes. In case the water source selected for artificial recharge has an unfavourable chemical or bacteriological composition, then chemical precipitation and clogging may take place in recharge ponds, infiltration ditches, at injection wells, in the groundwater system itself or at abstraction wells. Precipitation may include deposition of calcium carbonate, iron hydroxides and the formation of bacterial sludge. Precipitates affect the flow of water from the infiltration areas to the abstraction wells. Another factor that plays a role is the silt content of surface water that may seal the bottoms of recharge ponds, ditches and the screens of injection wells. In many places these problems can be avoided by introducing a pre-treatment phase before the water is applied in artificial recharge schemes.

6.3.2 *Groundwater demand*

Quantitative aspects

Groundwater demand studies are usually not dealt with in textbooks on hydrogeology. The topic is usually discussed in literature that is meant for water resources managers, water engineers and civil or agricultural engineers. However, when undertaking a groundwater availability study, one will have to be aware of the water demand issues in the area concerned (see section 6.1.1). Studies will

have to be set up to obtain an idea on the amount and quality of the groundwater that is required in the future. These issues can be further elaborated if one first realises for which purpose groundwater is used:
– Domestic and public consumption in towns and metropolitan areas
– Domestic and livestock consumption in rural areas
– Agricultural consumption in irrigation schemes
– Industrial use in processing and cooling
– Commercial use in bottled mineral water

The future water demand for domestic and public consumption can be estimated from existing population numbers and growth rates. These numbers and rates can be used to calculate future population numbers. Multiplication of the future population with a so-called 'per capita' water consumption gives an impression of the water demand that can be anticipated. The per capita consumption is the entitled water use per inhabitant or per family and is often prescribed by the central government of a country.

Future water demand can also be determined using existing data on water consumption. Figure 6.11 presents an illustrative diagram outlining the use of existing water consumption data to predict the groundwater demand for the municipality of Kenhardt in South Africa (Ministry of Water Affairs, 1979). The diagram presents data on groundwater consumption, compiled from records over the period 1964 to 1977.

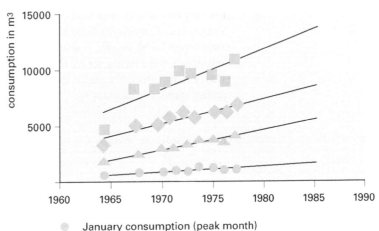

Figure 6.11. Chart showing municipal groundwater consumption plotted against time.

By extending the lines of existing trends the anticipated groundwater demands for 1985 were estimated. The figure shows that for 1985 the anticipated monthly peak demand would have been in the order of 18,000 m³, and the dry season demand in the period from October to April would have amounted to about 80,000 m³. Expressed as a total, the yearly consumptive demand for 1985 was anticipated at 140,000 m³.

Extending trends of groundwater consumption to the future can be carried out in the way illustrated in Figure 6.11. These are simply straightline extensions. Other methods of trend extension can be engaged in case it is indicated that the consumption pattern will change drastically in the future. Industrial development, for example, may be planned in a town leading to an industrial demand on groundwater, while the influx of factory labourers will also generate an additional domestic water demand. This could involve a step-wise or exponential increase in groundwater demand. On the other hand 'water saving measures' may be implemented, following a demand-oriented approach (see Figure 6.2). Consensus is needed amongst water managers, water engineers and hydrogeologists on water demand issues in towns and rural areas.

Qualitative aspects

The discussion so far has centred on the quantitative aspects of groundwater demand. The other important aspect is water quality. Water quality criteria can be formulated for the various purposes groundwater may be used for. Different criteria apply for water that is used for domestic and public purposes, for irrigation and for industrial cooling and processing. These criteria are based on setting recommended and permissible levels to physical, chemical and bacteriological substances that may be present in groundwater. Table 6.5 shows permissible concentrations of, mostly, inorganic chemical constituents in water to be considered for domestic consumption (drinking water). These concentrations were partly set by the United States Public Health Service (Davis et al, 1967). Naturally, in other countries a different set of water quality standards may apply.

Table 6.5. Permissible (maximum recommended) concentrations in drinking water in mg/l.

Constituent	Concentration	Constituent	Concentration
Total dissolved solids	1500	Copper	1.0
Arsenic	0.05	Fluoride	1.5
Bicarbonate	500	Iron	1.0
Cadmium	0.01	Manganese	0.05
Calcium	200	Nitrate	20
Chloride	250	Sulfate	250
Chromium	0.05	Zinc	5

6.3.3 *Local groundwater development*

Optimum site selection Groundwater studies focusing on local groundwater development may be initiated after regional assessments have been made (sections 6.2.1 and 6.3.1). In these local studies one tries to give an answer to the following questions. Where are the optimum sites for the abstraction of groundwater, taking into account the presence of local aquifers and the related groundwater availability and quality? What will be the design of the water works and how much water can be abstracted at individual installations? These questions will be highlighted below in a step-wise fashion.

The determination of optimum sites for groundwater abstraction requires a good understanding of the dimensions and characteristics of the local aquifers and aquitards, and the local (fresh) groundwater balance. The interpretation of aerial photographs, hydrogeological mapping, the inventory of existing wells, and the execution of surface geophysical investigations, exploration drilling, and pumping tests, enhance this understanding (see Table 6.3).

For the actual selection of the optimum sites for abstractions (e.g. a well or well field), a number of guidelines is usually followed. These guidelines are summarised below and are also presented as questions in Figure 6.12:

i) *Minimising environmental and socio-economic damage.* For the installation of wells, these issues should be addressed to avoid harm to the natural environment, to prevent damage to existing water supplies, and to avoid undesired social and economical consequences to the local population.

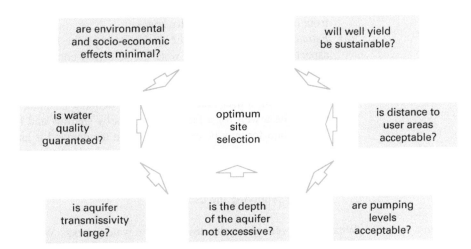

Figure 6.12. Diagram showing vital checks that are considered when siting a (production) well.

ii)   *Guaranteeing sustainable yield.* Although groundwater may be available on a regional scale (section 6.3.1), one could imagine that locally insufficient groundwater is available and sustainable yields to wells are not guaranteed. Decreasing yields could be witnessed at wells located in isolated local aquifers that receive little recharge.

iii) *Searching for high aquifer transmissivities.* Wells are preferably installed in aquifers with high transmissivities. Since transmissivity is the product of the coefficient of permeability and aquifer thickness, one will have to identify sites where sufficiently thick permeable rocks are present below the groundwater table.

iv)   *Taking into account the depth of the aquifer.* The deeper the aquifer is located below land surface, the higher will be the investment cost for well installation. On the other hand, the development of shallow aquifers may carry the risk of contamination. In case no protective aquicludes (e.g. clays, shales etc) are confining these aquifers, the risk of groundwater contamination from surface sources is imminent. The best choice would be to exploit aquifers that are not too deep and where the risk of contamination is minimal.

v)   *Considering groundwater levels.* Groundwater levels (pumping levels) that are too deep result in high operational costs. One should also realise that deep aquifers do not necessarily have deep pumping levels. As a result of their dynamic behaviour, levels associated with deep aquifers may be near, or even above land surface.

vi)   *Optimising water quality.* The groundwater quality at the well site needs to be considered and can be compared with water quality standards. One has to be aware of quality changes that may be invoked through future pumping practices. Water of an unacceptable quality may be attracted from saline or brackish 'pockets' or from other poor water quality zones.

vii) *Estimating distances to user areas.* Perhaps the most important aspect to be considered in well site selection concerns the local demand for groundwater and the distance between the well and user area. What is the use of having a 'perfect' well at more than 100 km from a user area? When water demand is small, say one has to site a well for a farm, then a nearby shallow and less productive aquifer may be exploited. When the demand is larger, for example, for a village or town, then one may also have to consider aquifers farther away. The larger demand then justifies the costs for the installation of the well and the pumping of water through a pipeline supplying the user area.

Groundwater abstraction works

What can be said about the design of the water works? When suitable sites for the abstraction of groundwater have been identified one may think in more detail about the design and construction of the ground water abstraction works. These may include 'captation' works at natural springs, wells or well fields equipped with casing and screen and re-inforced or unlined galleries.

Design drawings for a captation at a spring usually show a 'concrete wall' to create a reservoir from where the water can be led through pipes to distribution networks. Figure 6.13 presents a cross section through a spring captation showing the wall and the reservoir, as well as a concrete cover to protect the water source. The dam and reservoir guarantee that all the water emerging at the spring is collected, and storage is created for fluctuations in water demand. Protection zones are designed around the spring to prevent contamination of the ground-water system.

Well design diagrams show details on casings and screens, pumps and pump housing, well head protection, monitoring facilities, and if present, a gravel pack. For the preparation of well designs, insight into the groundwater system is required and field investigations are neces-sary. Important aspects to be considered for well design are the purpose of the well, the type of rock to be encountered at the selected site, and the anticipated depth and abstraction rate. With regard to purpose, wells are usually classified as follows:
– Hand dug wells
– Exploration wells
– Production wells
– Observation or monitoring wells

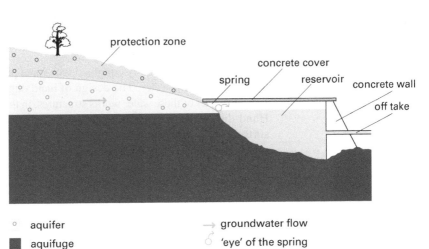

Figure 6.13. Cross section through a spring captation.

Figure 6.14.   View at a hand dug well in Orissa, India.

Hand dug wells are usually equipped with hand pumps to provide small communities with water. The design of hand dug wells is based on the large diameter of these water works structures, in the range of 1.4 to 2.0 m. Hand dug wells are dug in unconsolidated rock by shovels and all sorts of excavating equipment to shallow depths. To prevent caving the walls of these wells are re-inforced with concrete rings or other suitable material. Figure 6.14 shows a well, lined with stone blocks. Water entry into these wells at aquiferous zones is usually through porous cement rings or through a sand and gravel bed at the bottom of the well.

Exploration wells are drilled for investigation purposes (see section 2.3) and production or abstraction wells secure the supply for domestic, agricultural and industrial use. Exploration and production wells may show similar designs. Design drawings of these wells show that they have much smaller diameters than hand dug wells. Diameters in the order of 0.2 to 0.5 m are common. Figure 6.15 shows that in consolidated hard rocks, well design allows for an 'open hole' section. Groundwater can seep directly into the hole, emerging from open spaces in the rock aquifer. The design diagram for this well shows that any loose rock or unconsolidated silt and clay above the 'open hole' section is cased.

In unconsolidated rocks, a casing and screen design can be carried out in a variety of ways. Figure 6.15 shows that a single diameter casing-screen assembly can be designed. On the other hand, a telescoping construction can be installed or a set up using a reducer can be used. The well screens are placed opposite the most permeable parts in the aquifer. Depending on aquifer grain size, a gravel pack may or may not be placed. Particular attention in well design should also be paid to groundwater quality aspects. One does not drill holes into, and place screens at permeable zones with unsuitable water quality.

Design diagrams of observation wells for groundwater level or groundwater quality data collection programmes and monitoring purposes show that well diameters are usually smaller than the diameters for exploration or production wells. Diameters in the order of 0.05 to 0.2 m

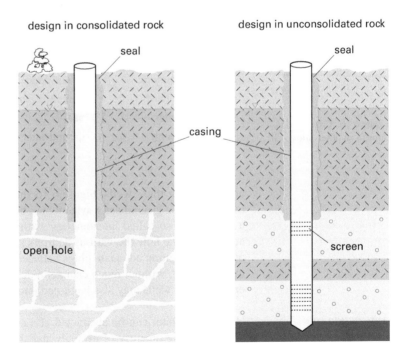

Figure 6.15. Diagrams of small diameter exploration and production wells.

○   sandy aquifer   clayey aquitard

   hard rock aquifer   ▽   groundwater table

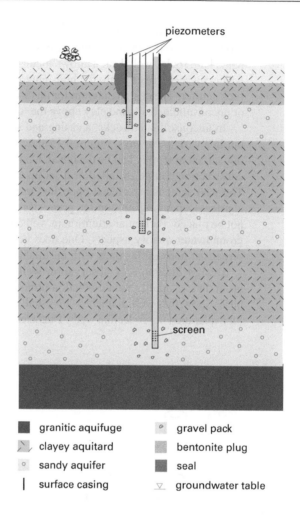

piezometers

screen

granitic aquifuge

clayey aquitard

sandy aquifer

surface casing

gravel pack

bentonite plug

seal

groundwater table

Figure 6.16. Cross section through an observation well.

are normal for these wells. Figure 6.16 shows that parameters at various depths in a groundwater system may be recorded by placing inside a single well, so-called 'piezometers' with screens placed at individual aquifers. In the hole the various aquifers should be isolated from each other by designing adequate plugs between the screens. Finalising one may note that, without much change in design, abandoned exploration or production wells may also be turned into observation wells.

Abstraction rate

How much water can be abstracted at a selected well site? To obtain an idea on the quantity of groundwater that could be abstracted, one usually needs interpreted data of surface geophysical investigations, and exploration drilling and pumping test programmes. In addition, one may require data from nearby well inventories, groundwater level data collection programmes, and well design diagrams. On the basis of these data one is able to select a maximum abstraction rate by considering the 'total drawdowns' in the well. The total drawdown can be defined as 'the total lowering of the groundwater level in the well with respect to an original groundwater level before the well is installed'. This drawdown is composed of:

- *'Aquifer loss':* Drawdown caused by the 'resistance to flow' in the idealized pumped aquifer (see section 3.2.3).
- *'Well loss':* Drawdown due to the 'resistance to flow' at the well screen and due to 'turbulence' inside the well.
- *'Losses due to partial penetration':* An extra lowering of the water level may be invoked in case a well screen does not fully penetrate an aquifer.
- *'Interference':* Drawdown caused by the effect of abstractions from other wells in the neighbourhood.
- *'Seasonal fluctuations':* In particular during the dry seasons an additional lowering of the groundwater level in the well may be expected.

The most important component of the total drawdown is aquifer loss. This component, which will be further elaborated in this textbook, can be estimated by applying the correct flow formula. For example, consider that the well will be constructed in a confined aquifer. For this case well formula (3.72) is an appropriate formula for steady flow describing the groundwater levels (hydraulic heads) around the well. If the transmissivity is known from pumping tests, then the re-arranged well formula can be used to compute the aquifer losses at the well, for the series of abstraction rates:

$$s = \phi_0 - \phi_w = -\frac{Q_o}{2\pi K_r H} \ln\frac{r_w}{R} \tag{6.3}$$

where:
s = aquifer loss at the well (m)
ϕ_w = groundwater level at the well during pumping (m)
ϕ_0 = original groundwater level before pumping (m)
Q_o = abstraction rate (m³/day)
K_r = coefficient of permeability (m/day)
H = aquifer thickness (m)
r_w = effective well radius (m)
R = distance to recharge boundary (m)

How does one proceed with the estimation of the maximum abstraction rate for the well? Assume that total drawdowns can be computed on the basis of flow formulae (see above) and other methods (Driscoll, 1986). Also note that the maximum abstraction rate (well yield) for the well can be defined as 'the abstraction rate giving an acceptable total drawdown at the well face'. Most organisations maintain a set of rules for the acceptable total drawdown in a well. Figure 6.17, illustrating a commonly used set of rules, shows that total drawdowns should be such that:

i) *For confined aquifers:* Groundwater levels (pumping levels) in a well are not lowered below the top of the aquifer.

ii) *For unconfined aquifers:* Groundwater levels in a well which fully penetrates the aquifer, should not drop below a level corresponding with 30 to 60% of the saturated aquifer thickness.

iii) *For aquifers pumped by suction pumps:* Groundwater levels should not drop below the maximum suction lifts of these pumps. This lift is usually in the order of 6 to 7 m.

The maximum abstraction rate can now simply be estimated by comparing, for a series of selected abstraction rates, the corresponding total drawdowns, with the rules for acceptable drawdowns. From the series, the maximum abstraction rate with an acceptable total drawdown can be selected.

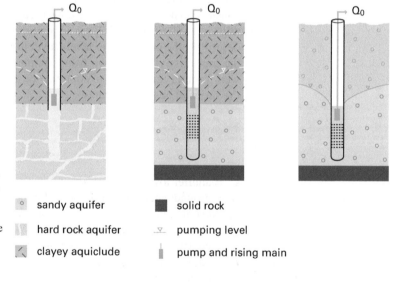

Figure 6.17. Minimum pumping levels in wells for the confined case in hard rock (left), the confined case in unconsolidated rock (middle) and the unconfined case in unconsolidated rock (right).

∘ sandy aquifer		▪ solid rock	
hard rock aquifer		▽ pumping level	
clayey aquiclude		pump and rising main	

Exercises

Exercise 1: Hydrogeology and rocktypes

In an area investigations including hydrogeological mapping have been carried out. The following rock types, illustrated in cross sections in Figure E.1, have been identified: i) partly cemented sandstone, ii) buried riverbed deposits of coarse sandy material, and iii) dense metamorphic schists and phyllites with very little joints and faults.

Assignment 1
Describe briefly where open spaces in each of the three rock types may occur.

Assignment 2
Which type of porosity will be observed at each of the rock types?

Assignment 3
Which rock type normally has the largest and which rock has the lowest permeability?

Assignment 4
What are common ranges for the coefficients of permeabilities for each of the rock types?

sandstone   river bed deposits   schists and phyllites

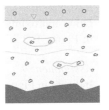

Figure E1. Cross-sectional sketches of the rock types in the investigation area.

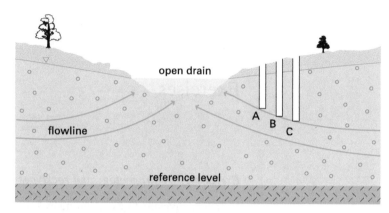

Figure E.2. Vertical section showing groundwater flow to the drain (not to scale).

Exercise 2: The concept of hydraulic head

Consider groundwater flow in an unconfined aquifer towards an open drain with a constant open water level. Figure E.2 shows a vertical section across the drain indicating flow lines and the position of the groundwater table. There is no recharge and discharge at the groundwater table. At the three wells shown in the section with screens at A, B and C, the following elevation heads and pressure heads are given:

Screen	Elevation head	Pressure head
A	10 m	5 m
B	8 m	8 m
C	7 m	10 m

Assignment 1
Determine the hydraulic heads at A, B and C.

Assignment 2
What is the status of the groundwater table: a hydraulic head contour line or a flow line? Please explain the answer.

Assignment 3
Draw approximately the hydraulic head contour lines of 15, 16 and 17 m in Figure E.2.

Assignment 4
Indicate in Figure E.2 the elevation heads and pressure heads at A, B, C.

Note: Use for the elevation heads and pressure heads respectively the symbols z_A, z_B, z_C and $h_{p(A)}$, $h_{p(B)}$, $h_{p(C)}$. Clearly show the pressure heads relative to each other, and to the groundwater table shown in the section.

Exercise 3: Computation of regional groundwater flow

Consider the extensive alluvial plain shown in Figure E.3. The plain is bordered by marly limestones in the west and east, and by a sea coast in the south. In the north, the alluvial plain narrows, but also extends into an adjacent plain. Underlying the plain is an extensive groundwater system consisting of a single sandy aquifer overlain by a less permeable top layer. The marly limestones that also form the basis of the aquifer are considered to be impermeable.

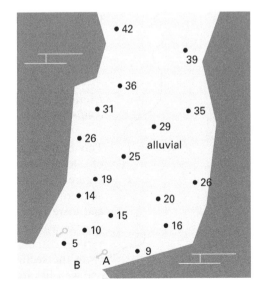

- ▽ groundwater table
- ○ sandy aquifer
- ○ top layer
- ▯ marly limestone
- ⌐ spring

Figure E.3. Map (below) showing average groundwater levels in the aquifer and section (above), showing the main flow components.

In the aquifer, the mainly horizontal groundwater flow is directed from the northern boundary of the plain towards the coast. With the exception of section AB where groundwater discharges, the shown stream has no hydraulic connection with the aquifer. Before groundwater abstractions were put into place, data on groundwater levels (hydraulic heads) were collected in exploration wells that were turned into observation wells. Figure E.3 shows the average groundwater levels of several observation years.

Pumping tests were also carried out in the exploration wells. In the northern part of the plain these tests indicated an average aquifer transmissivity of 2500 m^2/day. Aquifer material in the southern part of the plain is rather fine-grained and the tests indicated transmissivity values in the order of 1500 m^2/day.

Assignment 1
Draw groundwater level contour lines (groundwater head contour lines) for the alluvial aquifer, using a suitable contour line interval.

Assignment 2
Compute the groundwater flow rate through the aquifer at the entrance of the plain, using the appropriate formulation of the Darcy's Law equation and contour lines in the range of 35 and 40 m.

Assignment 3
Also compute the groundwater flow rate through the aquifer near the coast, using contour lines in the range of 10 and 15 m.

Assignment 4
Compare the groundwater flow rate through the aquifer at the entrance of the plain with the flow rate near the coast, and explain any differences in rate.

Exercise 4: Groundwater flow in a groundwater system with a semi-confining layer

This exercise deals with groundwater flow in a system composed of an unconfined and semi-confined aquifer. The section in Figure E.4 shows the aquifers and the separating semi-confining aquitard. The flow in the aquifers can be considered as horizontal and there is no significant flow component at an angle to the section shown. In the semi-confining layer the flow can be considered vertical.

In the section in Figure E.4 two observation wells are shown; one well has its screen at location, e, in the unconfined aquifer and

cross section with shallow and deep well

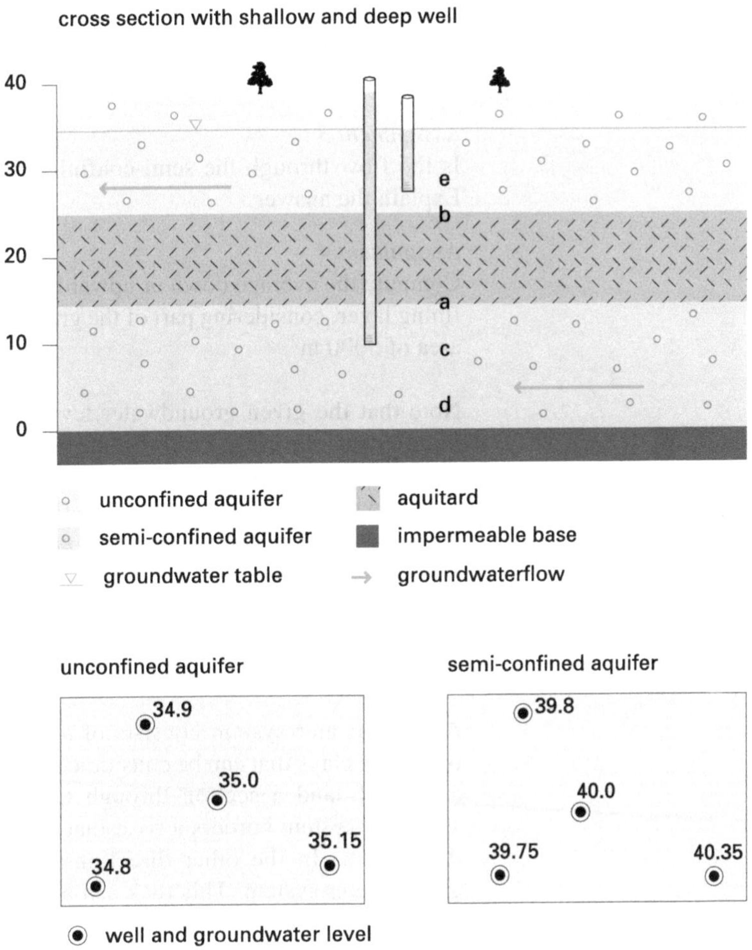

Figure E.4. Section and maps showing respectively flow and groundwater levels in a semi-confined groundwater system.

the other well is screened at c, in the semi-confined aquifer. The groundwater level measured in the well in the unconfined aquifer is at 35 m above the top of an impermeable base that can be considered as a reference level. The groundwater level measured in the well protruding into the semi-confined aquifer is at 40 m above reference.

Pumping tests have been carried out in the area and indicated an average transmissivity of 1000 m^2/day for the semi-confined aquifer and an average vertical resistance for the semi-confining layer in the order of 500 days.

Assignment 1
What are the verticals eb and acd?

Assignment 2
Determine the groundwater levels (hydraulic heads) at the other points along the section at a, b, and d.

Assignment 3
Is the flow through the semi-confining layer downward or upward? Explain the answer.

Assignment 4
Compute the average down or upward flow rate through the semi-confining layer, considering part of the groundwater system with a surface area of 5000 m^2.

Note that the given groundwater levels at the observation wells are representative average levels for the groundwater system to be considered. As an illustration, maps of the aquifers are presented in Figure E.4 showing the groundwater levels (hydraulic heads) that could have been used to assess representative groundwater levels for flow rate calculations.

Exercise 5: Groundwater modelling

A groundwater system consists of a very coarse sandy aquifer and overlying clays that can be considered an aquiclude. Figure E.5 shows a map of- and a section through the groundwater system. In the south the system borders a river that is in full hydraulic contact with the aquifer. In the other directions, quartzitic rock surrounds the groundwater system. This rock can be considered as an impermeable aquifuge. It is further given that in view of the impermeable nature of the overlying aquiclude, no direct recharge takes place into the aquifer.
A large well field consisting of several pumping wells is located in the centre of the system and abstracts groundwater from the aquifer. The well field is managed by a local water supply company. The company wants to know to what extent the groundwater levels (hydraulic heads) in the aquifer will decrease in case the existing abstraction rate at the well field, i.e. 12,500 m^3/day is doubled to 25,000 m^3/day. This may affect a small number of private wells in the surroundings that pump very small amounts of water for domestic use.
    The water supply company intends to set up a simple groundwater flow model for the groundwater system in order to obtain a first estimate on the effect of increased abstractions on the groundwater levels. The hydrogeologist who has to carry out this task is requested to perform the following assignments:

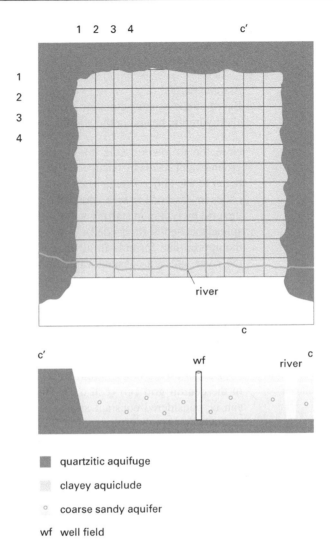

Figure E.5. Map and section of the modelled groundwater system.

quartzitic aquifuge

clayey aquiclude

coarse sandy aquifer

wf well field

Assignment 1
Prepare the conceptual model for the groundwater system.

Assignment 2
Set up the model grid by making an enlargement of the shown grid on a separate piece of paper, complete column and row numbers, and add any other relevant information.

Note that for this simple model it is decided to take a grid consisting of squares with a cell size of 1 * 1 km. The area to be modeled is 11 * 11 km and includes the river that has a width of approximately 100 m. The grid can be drawn in such a way that the river coincides with the centers of the 11th row of cells.

Assignment 3
Based on the existing situation, compute steady state groundwater levels (hydraulic heads) in the aquifer.

Note the following. One could carry out the computations of the groundwater levels by hand, cell by cell, but this would take time. It is better to use either a spreadsheet or a full groundwater flow modelling code. In case one uses a spreadsheet proceed as follows: 1) Define the grid by typing in starting values for the groundwater levels (11*11 spreadsheet cells are needed); 2) set recalculation to manual; 3) insert the appropriate formula in all model cells except the cells covering the river where a constant value is maintained; and 4) compute the groundwater levels, iteration by iteration, using the calculation function of the spreadsheet.

Hint: Consider for the appropriate formulation of the formulae, the equations introduced in section 3.2.2 on flow modelling. Pay particular attention to the formulae to be inserted in the boundary cells and the cell with the well field.

The following information is also given. The well field is placed in the central grid cell (cell 6,5). The open water levels in the river can be considered constant at +30 m and the starting values for the groundwater levels of the aquifer may be assigned the same value. It is further given that the aquifer has a constant thickness of 12.5 m. The coefficient of permeability of the aquifer has only been roughly estimated at 150 m/day.

Assignment 4
Calibrate the groundwater flow model and mention the groundwater level in the cell with the well field.

Note the following. Proceed as follows: 1) select the most likely parameter to be calibrated; 2) change the value of this parameter until the differences between computed and observed groundwater levels are minimal.

The following additional information is given. The observed groundwater level at cell (6,6) is +24.00 m and at cell (6,7) the level is +25.70 m. Finally, at cells (3,4) and (10,9), the observed groundwater levels are respectively at +24.85 m and at +28.30 m.

Assignment 5
Compute the decrease in groundwater levels at all the model cells when the abstraction rate is doubled.

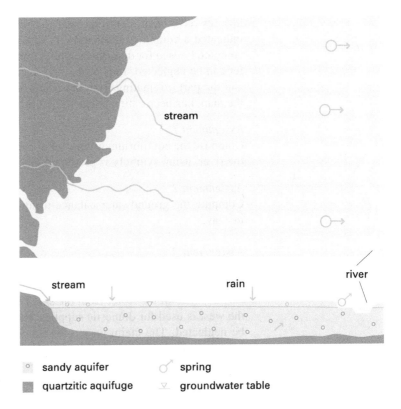

Figure E6. Map (above) and section (below) showing the alluvial plain between quartzite mountains and a river.

- ○ sandy aquifer
- ▓ quartzitic aquifuge
- ♂ spring
- ▽ groundwater table

Assignment 6

What is the decrease of the groundwater level in the cell where the well field is installed?

Exercise 6: The groundwater balance

Consider the alluvial plain shown in Figure E.6. The plain is bordered by a mountain range in the west. The range consists of impermeable granite and quartzite. The plain is bordered in the east by a large river. The shown plain between the mountains and the river has a surface area of 33.7 km^2 and is underlain by a groundwater system consisting of sandy deposits. These deposits can be considered as a single unconfined aquifer.

The aquifer is in a state of equilibrium; i.e. in the long term there is no increase or reduction in storage. The aquifer receives recharge from precipitation in the order of 130 mm/year. A number of small streams which enter the sandy deposits at the foot of the mountain range also recharge the groundwater system at an estimated rate of 3000 m³/day. Springs have been identified near the main river and measurements indicated a combined discharge of 5000 m³/day. The spring water is completely used for domestic supplies and any return flow to the aquifer can be neglected. In addition, the river is known to have baseflow and the part originating from the area west of the river and shown on the map, has been estimated at 10,000 m³/day.

Assignment 1
Compose the equilibrium groundwater balance for the aquifer, west of the river, using symbols as presented in section 4.1.

Assignment 2
Compute the groundwater balance using units of i) mm/year and ii) m³/day.

Assignment 3
Does the balance balance and if not what could be the reason?

Assume that all the water from the small streams entering the plain in the west is used for domestic supplies. Return flow to the aquifer can be neglected. The claim on stream water for domestic consumption affects the flow at the springs.

Assignment 4
Compose and compute the new groundwater balance after a new state of equilibrium has been reached.

Assignment 5
In the transition to the new state of equilibrium will there be an increase or decrease in groundwater storage? Explain your answer.

Exercise 7: Groundwater chemistry

The idea of this exercise is to follow up changes in chemical composition of recharge water from precipitation, which enters an unsaturated zone and then flows through a saturated groundwater system. Some simple processes affecting the chemistry of the recharge water will be dealt with. Redox reactions and ion exchange will be neglected in this exercise.

Imagine that the area concerned consists of unconsolidated rock mainly made up of sands and some peat layers. The soil contains a certain amount of organic matter. The sand in the unsaturated zone and

Table E.1. (Electro-neutral) concentrations of chemicals in precipitation.

Chemical	Concentration (mg/l)	Chemical	Concentration (mg/l)
Chloride	8	Sodium	4
Sulfate	3	Potassium	1
Nitrate	1	Calcium	2
Bicarbonate	3	Magnesium	0.5
Carbon dioxide	10		
(pH)	(5.6)		

Table E.2. Chemical concentrations of anthropogenic substances.

Chemical	Concentration (mg/l)	Chemical	Concentration (mg/l)
Chloride	18	Sodium	15
Sulfate	8	Potassium	5
Nitrate	6		

in the saturated groundwater system contains considerable amounts of calcite ($CaCO_3$). The release of anthropogenic substances is known to take place in the unsaturated zone.

The precipitation entering the unsaturated zone has a chemical composition as presented in Table E.1. Through decay of organic material, 22 mg/l of carbon dioxide (CO_2) is freed in the soil. The total amount of carbon dioxide in the recharge water is increased to 32 mg/l, and 15 mg/l carbon dioxide is subsequently used in the dissolution of calcite.

Assignment 1
Compute the new chemical composition of the recharge water after dissolution.

An increase in the concentration of total dissolved solids (TDS) in the recharge water is caused by evapotranspiration in the unsaturated zone. Assume that the uptake of dissolved solids by plant roots can be neglected and that 2/3 of the recharge water is lost by evapotranspiration.

Assignment 2
Compute the new chemical composition of the recharge water after the effect of evapotranspiration.

The down flowing recharge water is mixed with anthropogenic substances in the unsaturated zone. Assume that the introduction of the substances does not trigger off substantial chemical reactions. The concentrations of the anthropogenic components are shown in Table E.2.

Assignment 3
Determine the chemical composition of the recharge water after it has passed through the zone with the anthropogenic substances.

The dissolution of calcite continues in the saturated groundwater system. Assume that during the dissolution of calcite the concentration of carbon dioxide decreases to 12 mg/l.

Assignment 4
Compute the final chemical composition of the recharge water after dissolution in the saturated groundwater system.

For completion of the exercise the following information is given: Atomic weights: Cl = 35.5; S = 32; O = 16; N = 14; H = 1; C = 12; Na = 23; K = 39; Ca = 40; Mg = 24.

Exercise 8: Groundwater availability

Consider the alluvial plain of exercise 3 again. Terms of the groundwater balance and the water balance for the root zone have been identified for the plain. The terms can be described as follows.

The mainly horizontal groundwater inflow at the northern entrance of the plain (Q_{lsi}) is directed towards the sea where groundwater outflow takes place ($Q_{surfout}$). At the southern side of the plain part of the groundwater discharge is at springs (Q_{spring}) and at section AB of the river ($Q_{surfout}$).

Recharge into the less permeable top layer and the underlying aquifer originates from precipitation (P). Part of this precipitation is lost by evapotranspiration (E) above land surface and from the root zone. The surface runoff (R) in the area itself is small and will further be neglected. There is also recharge from surface water irrigation (Q_{irr-s}) which is evenly distributed over the area. Note that all baseflows and flood waters entering the area via the river shown in Figure E.3 are in fact used for irrigation. The surface area of the alluvial plain area amounts to 50 km².

Assignment 1
Compose the long term equilibrium water balance for the root zone considering recharge from precipitation. Assume that there are hardly areas with shallow water tables. Use the symbols introduced in section 4.2.

Assignment 2
Also set up the equilibrium groundwater balance for the saturated groundwater system, using the symbols in section 4.1.

Assignment 3
Indicate the terms of the groundwater balance on Figure E.3.

Field investigations have revealed that the terms of the water balance for the root zone and the groundwater balance for the saturated groundwater system can be estimated as follows:

Term		Quantity
P	=	500 mm/year
E	=	65% of P
Q_{lsi}	=	25,000 m³/day
Q_{irr-s}	=	105 mm/year
Q_{spring}	=	19,500 m³/day
$Q_{surfout}$	=	19,500 m³/day (river at AB)
$Q_{surfout}$	=	24,000 m³/day (sea)

Assignment 4
In case a 'safe yield' policy is followed, determine the potential groundwater availability in m³/day.

Assignment 5
Mention a modern technique that one can use for the verification (checking) of the above terms, which are based on field measurements?

On the basis of the computed value for the potential groundwater availability, an optimum groundwater availability of 40,000 m³/day has been recommended. It is decided that groundwater abstractions to this amount will be supplied for irrigation to complement the existing supply from surface water. During irrigation practices 75% of the groundwater supplied is used by crops and eventually lost as transpiration. The other 25% returns to the saturated part of the top layer and aquifer (Q_{irr-g}). Assume that the precipitation rate (P), the percentage of the precipitation lost by evapotranspiration (E), the recharge from irrigation by surface water (Q_{irr-s}), and the subsurface groundwater inflow at the northern boundary (Q_{lsi}) remain unaltered by the introduction of the groundwater abstractions.

Assignment 6
Compose the long term equilibrium groundwater balance again, taking into account the information above.

Assignment 7
Also, compute the new combined value for the groundwater discharge to surface water ($Q_{surfout}$) and to the springs (Q_{spring}) in m³/day.

Assignment 8
What is the difference between these values, and the values for these parameters before the abstractions were put into place?

Consider the dry season when there is no precipitation, and irrigation by surface water. It is further known that in the dry season the abstraction rates are at 60,000 m³/day. Return flow remains as specified: 25%. In the dry season the average subsurface inflow (Q_{lsi}), the discharge to surface water $(Q_{surfout})$ and the flow towards the springs (Q_{spring}) decrease to:

Term		Quantity
Q_{lsi}	=	15,000 m³/day
$Q_{surfout}$	=	1500 m³/day
Q_{spring}	=	500 m³/day

Assignment 9
Compose the groundwater balance for the dry season.

Assignment 10
Determine the amount taken from groundwater storage (S_{bal}) in m³/day.

Assignment 11
Find an approximate formula to calculate the average lowering of the groundwater table in the dry season considering the (average seasonal) amount of water taken from storage, the duration of the dry season, the specific yield of the top layer, and the surface area.

Assume that the average specific yield of the top layer is 0.2 and that a dry season lasts for 150 days.

Assignment 12
Use the derived formula to compute the average lowering of the groundwater table in the alluvial plain at the end of the dry season.

Selected answers to exercises

1.2 Sandstone: Primary and secondary porosity
Buried river bed deposits: Primary porosity
Schists and phyllites: Secondary porosity

1.4 Sandstone: 0.1–10 m/day
Buried river bed deposits: 10–200 m/day
Schists and phyllites: 0.00001–0.5 m/day

2.1 Point A: 15 m
Point B: 16 m
Point C: 17 m

2.2 Groundwater table is a flow line.

3.2 Q in the range of 25,000 to 30,000 m^3/day

3.3 Q in the range of 40,000 to 45,000 m^3/day

4.2 Total head at a: 40 m
Total head at b: 35 m
Total head at d: 40 m

4.4 $Q_z = 50$ m^3/day

5.4 Calibrated groundwater level in well field cell: About 21.10 m

5.6 Decrease in groundwater level in well field cell: About 8.90 m

6.2 162.5 mm/year ~ 162.45 mm/year (equilibrium is indicated)
15,002 m^3/day ~ 15,000 m^3/day (equilibrium is indicated)

6.4 130.0 mm/year ~ 130.3 mm/year (equilibrium is indicated)
12,002 m^3/day ~ 12,000 m^3/day (equilibrium is indicated)

7.1   Increase in Ca^{2+}: 13.8 mg/l
      Increase in HCO_3^-: 41.6 mg/l

7.2   Increase by evapotranspiration:
      Concentration recharge water: Concentration precipitation water
      = 3:1

7.4   Increase in Ca^{2+}: 35.4 mg/l
      Increase in HCO_3^-: 108.1 mg/l

8.2   Groundwater balance: $[Q_{prec} + Q_{irr-s} + Q_{lsi}] - [Q_{surfout} + Q_{spring}]$
      $= 0$

8.4   Potential groundwater availability: 63,356 (about 63,000) m^3/day

8.8   Reductions: 29,644 m^3/day

8.10 From storage: 32,000 m^3/day

8.12 Groundwater table decrease: 0.48 m

References and Bibliography

Abushaar R.F. 1997. *Design of an artificial recharge system through groundwater modelling.* Delft: (Unpublished) MSc thesis HH 362 for IHE.

Anderson M.P. & Woessner W.W. 1992. *Applied Groundwater Modelling.* London: Academic Press Inc/Harcourt Brace Jovanovich Publishers.

Anonymous 1975. *Water: Origins and demand, conservation and abstraction.* Environmental Control and Public Health Units 3-4. Milton Keynes: The Open University Press.

Appelo C.A.J. & Postma D. 2009. *Geochemistry, groundwater and pollution.* Rotterdam: A.A. Balkema.

Athavale R.N., Chand R. & Rangarajan R. 1983. Groundwater recharge estimation for two basins in the Deccan Trap Basalt Formation. *Hydrological Sciences Journal*, 28: 525–538.

Badon Ghijben W. 1888. Nota in verband met de voorgenomen putboring nabij Amsterdam (in Dutch). *Tijdschrift van het Koninklijk Instituut van Ingenieurs*, 1888–1889: 8–22.

Barry R.G. 1971. *Introduction to physical hydrology.* London: Methuen.

Bear J. 1972. *Dynamics of fluids in porous media.* New York: Elsevier Science.

Bear J. & Verruijt A. 1987. *Modelling groundwater flow and pollution.* Boston: Reidel Publishing Company.

Bisht S.B. 1999. *Modeling anisotropy in the Hondsrug area of The Netherlands.* Delft: (Unpublished) MSc thesis HH 361 for IHE.

Bisson R.A. & Lehr J.H. 2004. *Modern groundwater exploration.* Hoboken: John Wiley and Sons.

Bradley E. & Phadtare P.H. 1989. Paleo-hydrology affecting recharge to over-exploited semi-confined aquifers in the Mehsana area. *Journal of Hydrology*, 108: 309–322.

Brassington R. 1998. *Field hydrogeology.* Chicester: John Wiley

Campbell M.D. 1973. *Water well technology: field principles of exploration, drilling and development of groundwater and other selected minerals.* New York: McGraw-Hill.

CGWB (Central Groundwater Board of India) 1989. *Various maps and water balance data for the Faridabad area.* New Delhi: (Unpublished) material of the North Western Region office.

Charbeneau R.J., Bedient P.B. & Loehr R.C. 1992. *Groundwater remediation.* Water quality management Library. Lancaster: Technomic Publishing Company.

Cherry J.A. & Freeze R.A. 1979. *Groundwater.* Englewood Cliffs: Prentice Hall.

Chiang W. & Kinzelbach W. 2001. *3D-Groundwater modeling with PMWIN.* Berlin/Heidelberg: Springer Verlag.

Darcy H. 1856. *Les fontaines publiques de la ville de Dijon.* Paris: V. Dalmont.

Davis S.N. & De Wiest R.J.M. 1967. *Hydrogeology.* New York/London/Sydney: John Wiley and Sons.

De Laat P.J.M. 1980. *A model for unsaturated flow above a shallow water-table, applied to a regional subsurface flow problem.* Thesis. Agric. Res. Rep. 895. Wageningen: Pudoc.

De Marsily G. 1986. *Quantitative hydrogeology: groundwater hydrology for engineers.* New York: Academic Press.

Dupuit J. 1863. *Etudes theoriques et pratiques sur le mouvement des eaux dans les canaux decouverts et a travers des terrains permeables* (in French). Paris: Dunod, 2nd edition.

De Wiest R.J.M. 1965. *Geohydrology.* New York: John Wiley.

De Wiest R.J.M. 1969. *Flow through porous media.* New York: Academic Press.

Domenico P.A. 1972. *Concepts and models in groundwater hydrology.* New York: McGraw-Hill.

Doorenbos J. & Pruitt W.O. 1977. *Crop water requirements.* Rome: FAO Irrigation and Drainage Paper 24.

Driscoll F.G. 1986. *Groundwater and wells.* St Paul, Minnesota, USA: Johnson Division.

Dufour F.C. 2000. *Groundwater in The Netherlands: Facts and figures.* Utrecht: TNO-NITG.

Euroconsult 1986. *Feasibility study of the Ivancho-Pedernalis area.* Arnhem: (unpublished) report to the Government of the Dominican Republic.

Euroconsult 1988. *Baluchistan groundwater and trickle irrigation project.* Arnhem: Proposal for technical consultancy services to the Government of Pakistan.

Euroconsult 1989. *Projet de mise en valeur agricole de la Basse Plaine des Gonaives, Haiti* (in French). Arnhem: (unpublished) report to GTZ and the Government of Haiti.

Fetter C.W. 2000. *Applied hydrogeology.* Englewood Cliffs: Prentice Hall.

Fetter C.W. 2000. *Solution manual to accompany Applied hydrogeology.* Englewood Cliffs: Prentice Hall.

Fried J.J. 1975. *Groundwater pollution: theory, methodology, modelling and practical rules*. Amsterdam: Elsevier.

Fitts C.R. 2002. *Groundwater Science*. San Diego: Academic Press.

Gehrels J.C. 1999. *Groundwater level fluctuations*. Thesis. Utrecht: Elinkwijk.

Geochem 1990. *Herkomst van de verhoogde arseengehaltes in de bodem van De Huet, Doetinchem* (in Dutch). Utrecht: (Unpublished) report to the Municipality of Doetinchem.

Geochem 1992. *Verhoogde arseen concentraties in de bodem van het CIOS terrein in Haarlem* (in Dutch). Utrecht: (Unpublished) report to the Municipality of Haarlem.

Gieske A.S.M. 1992. *Dynamics of groundwater recharge; a case study in semi-arid Botswana*. PhD thesis for the Free University of Amsterdam. Enschede: Febodruk BV.

Gonzales A.L., Nonner J.C., Heijkers J. & Uhlenbrook S. 2009. *Comparison of different base flow separation methods in a lowland catchment*. Hydrology and Earth System Sciences. Copernicus Publications.

Hamill L. & Bell F.G. 1986. *Groundwater resources development*. London: Butterworth.

Hem D.J. 1959. *Study and interpretation of the chemical characteristics of natural water*. Washington: Geological Survey Water-supply paper 1473.

Hemker C.J. & Van Elburg H. 1987. *Micro-Fem Version 2.0 User's Manual*. Amsterdam: published by the authors.

Herzberg A. 1901. Die Wasserversorgung einiger Nordseebader (in German). *Zeitung fur Gasbeleuchtung und Wasserversorgerung*, 44: 815–819, 842–844.

Houston J. 1988. Rainfall–runoff–recharge relationships in the basement rocks of Zimbabwe. Estimation of natural groundwater recharge. *Mathematical and Physical Sciences*, Vol. 22. Boston: Reidel Publishing Company, pp. 349–365.

Huisman L. 1972. *Groundwater recovery*. London and Basingstoke: MacMillan.

Huisman L. & Olsthoorn Th. 1983. *Artificial groundwater recharge*. Boston: Pitman.

IHE 1992–2000. *Various groundwater modelling studies with MODFLOW*. Delft: (Unpublished) MSc theses HH153, HH245, HH274, and HH 324.

IHEG (Institute of Hydrogeology and Engineering Geology) 1988. *Hydrogeologic Map of China*. Chinese Academy of Geological Sciences. Hebei: China Cartographic Publishing House.

Ilaco 1990. *Groundwater model of the Rada Basin*. Arnhem: Rada integrated rural development project. (Unpublished) report to the Yemen Arab Republic.

Ipesa Consultores 1975. *Recargo de los aquiferos volcanicos del Rio Tepalcatepec* (in Spanish). Mexico DF: (Unpublished) report to the Government of Mexico.

Jacob C.E. 1950. *Flow of groundwater*. Engineering Hydraulics. New York: John Wiley and Sons.

Khan L.A. 1968. *Report on geohydrological investigations in the Bannu Basin, West Pakistan*. WAPDA (Water and Power Development Authority) Bulletin 15, and WASID Publication No 56.

Keller G.V. 1960. *Physical properties of the Oak Spring Formation, Nevada*. US Geol. Survey Prof. Paper 400-B, p. 396–400.

Kinzelbach W. 1986. *Groundwater modelling: An introduction with sample programs in BASIC*. New York: Elsevier Publishing Company.

Koefoed O. 1979. *Geosounding principles, 1*. Amsterdam: Elsevier Scientific Publishing Company.

Koudstaal R., Rijsberman F.R. & Savenije H.H.G. 1992. *Water and sustainable development*. Dublin: Proceedings International Conference on the Environment, Keynote papers 9/1–9/23.

Krauskopf K.B. 1967. *Introduction to geochemistry*. New York: McGraw-Hill Publishers.

Kruseman G.P. & De Ridder N.A. 1990. *Analysis and evaluation of pumping test data*. Wageningen: International Institute for Land Reclamation and Improvement.

Langmuir D. 1971. *Geochemistry of some carbonate waters in Central Pennsylvania*. Geochim. et Cosmochim. Acta, 35: 1023–1046.

Lvovich M.I. 1979. *World water resources and their future*. Translated by R.L. Nace, AGU.

Makkink G.F. & Van der Heemst M.D.J. 1965. *Calculation model for the actual evapotranspiration of cropped areas and other terms of the water balance equation*. Paris: Unesco Arid Zone Research Vol. 25.

Matthess G. 1982. *The properties of groundwater*. New York: John Wiley.

Mazor E. 1991. *Applied chemical and isotopic groundwater hydrology*. Buckingham: Open University Press.

McDonald M.C. & Harbaugh A.W. 1988. *MODFLOW, a modular three-dimensional finite-difference groundwater flow model*. Denver: Geological Survey Open File Report 91-536.

Mead D.W. 1919. *Hydrology*. New York: McGraw-Hill Book Company.

Meinzer O.E. 1923. *Outline of groundwater hydrology with definitions*. US Geol. Survey Water Supply Paper 577.

MGMR (Ministry of Geology and Mineral Resources) & TNO (Institute of Applied Geoscience). 1989. *Optimisation of the secondary groundwater level monitoring network for Zhengzhou City by application of the Kalman Filtering technique*. Delft: Report No 89-10 to the Government of China.

Ministry of Water Affairs 1977. *Groundwater investigation of the Dinokana area in Bophuta Tswana*. Pretoria: (unpublished) report to the Government of South Africa.

Ministry of Water Affairs 1979. *Groundwater resources investigation for Kenhardt Municipality in Cape Province.* Pretoria: (unpublished) report to the Government of South Africa.

Moench, A. 1994. *Specific yield as determined by type-curve analysis of aquifer test data.* Ground Water 32, no 6: 949–957.

Morris D.A. & Johnson A.I. 1967. *Summary of hydrological and physical properties of rock and soil materials, as analysed by the Hydrological Laboratory of the US Geological Survey 1948-60.* Geological Survey Water Supply Paper 1839-D PD1-D42, 42 pp.

Nath S.K., Patra H.P. & Shahid S. 2000. *Geophysical prospecting for groundwater.* Rotterdam: A.A. Balkema.

Nace R.L. 1960. *Water management, agriculture and groundwater supplies.* US Geol. Survey Circ. 415.

Nielson D.M. 1991. *Practical Handbook of Ground-Water monitoring.* Chelsea: Lewis Publishers.

Nonner, J.C. 1969. *Hydrogeologisch onderzoek van een gedeelte van het opvanggebied van de Rio di Fundres in de Zuid Alpen* (in Dutch). Amsterdam: (Unpublished) report to the Free University of Amsterdam.

NWASA (National Water and Sanitation Authority) & TNO (Institute of Applied Geoscience) 1996. *Evaluation of the effects of groundwater use on groundwater availability in the Sana'a Basin.* Sana'a/ Delft: Technical report No 5 Volume 1. (Unpublished) report to the Government of Yemen.

Olsthoorn T.N. 1985. The power of the electronic worksheet: modeling without special programs. *Ground Water* 27: 381–390.

Oude Essink G.H.P. 2001. *Density dependent groundwater flow—Salt water and heat transport.* IHE Lecture notes HH426/01/1.

Palmer C.M. 1992. *Principles of contaminant hydrogeology.* Boca Raton: Lewis Publishers.

Penman H.L. 1948. Natural evaporation from open water, bare soil and grass. *Proceedings of the Royal Society Ser. A*, 193: 120–145.

Perez, E.S. 1997. *Estimation of basin-wide recharge rates using spring flow, precipitation and temperature data.* Ground Water 35, no 6: 1085–1065.

Pollock D.W. 1989. *MODPATH Documentation of computer programs to compute and display pathlines using results from the U.S. Geological Survey modular three dimensional finite-difference groundwater model.* Reston: U.S. Geological Survey Open File Report 89–381.

Price M. 1996. *Introducing groundwater.* London: Chapman and Hall.

Rail Ch.D. 1989. *Groundwater contamination: Sources, control, and preventive measures.* Lancaster: Technomic Publishing.

RGD (Rijks Geologische Dienst) 1980. *Geologische Kaart van Nederland, Blad Heerlen* (in Dutch). Haarlem.

Ridder T.B. 1978. *Over de chemie van neerslag* (in Dutch). De Bilt: Royal Dutch Meteorological Institute, Report 78-4.

Senn A. 1946. *Geological investigations of the groundwater resources of Barbados*. B.W.I. Report for British Union Oil Company Limited.

Sibanda T., Nonner J.C. & Uhlenbrook S. 2009. *Comparison of groundwater recharge estimation methods for the semi-arid Nyamanhlovu area, Zimbabwe*. Hydrogeology Journal. Springer Verlag.

Sinha B.P.C. & Sharma S.K. 1988. Natural groundwater recharge estimation methods in India. Estimation of natural groundwater recharge. *Mathematical and Physical Sciences*, 222D. Boston: Reidel Publishing Company, pp. 301–311.

Shiklomanov I.A. 1997. *Comprehensive assessment of the freshwater resources of the world*. Stockholm: World Meteorological Organisation.

Strack O.D.L. 1989. *Groundwater mechanics*. Englewood Cliffs: Prentice Hall.

Stuyfzand P.J. 1989. *Hydrochemical evidence of fresh and salt water intrusion in the coastal dunes aquifer system in the Western Netherlands*. Gent: Proceedings of the 10th salt water intrusion meeting, pp. 9–29.

Stuyfzand P.J. 1999. Patterns in groundwater chemistry resulting from groundwater flow. *Hydrogeology Journal*, 7: 15–27.

Theiss C.V. 1935. The relation between the lowering of the piezometric surface and the rate and duration of discharge of a well using groundwater storage. *Trans. American Geophysical Union*, 16: 519–524.

Thiem C.V. 1906. *Hydrologische Methoden*. Leipzig: Gebhardt.

Times Books 1994. *Times Atlas of the World*. London.

Ting C. 1997. *Groundwater resources evaluation and management for Pingtung Plain, Taiwan*. PhD thesis, Amsterdam Free University.

Todd D.K. 1959. *Groundwater hydrology*. New York: John Wiley and Sons.

Topografische Dienst 1997. *Aerial photography of The Netherlands, mapsheet 40*. Emmen.

Toth J. 1962. A theory of groundwater motion in small drainage basins in central Alberta. *Journal of Geophysical Research*, 67: 4375–4387.

Toth J. 1963. A theoretical analyses of groundwater flow in small drainage basins. *Journal of Geophysical Research*, 68: 4795–4811.

Trottier J. & Slack P. 2002. *Managing water resources – Past and present*. Oxford: University Press.

Van Dam J.C. 1983. *Shape and position of the salt water wedge in coastal aquifers*. Hamburg: IAHS publication 146.

Van Elburg H., Engelen G.B. & Hemker C.J. 1989. *FLOWNET Users Manual*. Amsterdam: Free University of Amsterdam.

Vidanaarachchi C.K., Zhou Y. & Nonner J.C. 1998. *Optimisation of artificial recharge systems with infiltration galleries.* Proceedings of the third international symposium on artificial recharge of groundwater TISAR 1998. Rotterdam: Balkema Publishers, pp. 161–166.

Verruijt A. 1970. *Theory of groundwater flow.* London: MacMillan.

Walton W.C. 1970. *Groundwater resource evaluation.* New York: McGraw-Hill.

WAPDA (Water and Power Development Authority) & TNO (Institute of Applied Geoscience) 1983. *Technical report on groundwater resources in Domail Plain, Bannu and Karak Districts.* Peshawar/Delft: Report IV-1 to the Government of Pakistan.

Wilson E.M. 1983. *Engineering hydrology.* London: MacMillan.

Wood W.W. & Sanford W.E. 1995. *Chemical and isotopic methods for quantifying groundwater recharge in a regional, semi-arid environment.* Ground Water 33, no 3.

Wu Q. 1997. *Groundwater study in the Weerselo area, Overijssel, The Netherlands.* Delft: (Unpublished) MSc thesis HH 324 for IHE.

Yasin R.H. 1999. *Groundwater modelling of fractured karstic environments: A case study of the Eastern Aquifer Basin.* Delft: (Unpublished) MSc thesis 363 for IHE.

Zhang K. 1996. *Modelling tools for groundwater monitoring network design. Case study Valtherbos pumping station, Drenthe Province.* Delft: (unpublished) MSc thesis HH 274 for IHE.

Zheng C. 1990. *MT3D: A modular three dimensional transport model for simulation of advection, dispersion and chemical reactions of contaminants in groundwater system.* Report to the U.S. Environmental Protection Agency, Ada, OK.

Zheng C. & Bennett G.D. 1995. *Applied contaminant transport modelling: Theory and practice.* New York: Van Nostrand Reinhold.

Register